ASML and Dutch Physics

ASML and Dutch Physics

A History of the Advanced Research Center for
Nanolithography (ARCNL, 2014–2024)

Hein Brookhuis

Translated by
Erin Goedhart-Stallings

Amsterdam University Press

The publication of this book is made possible with financial support from ARCNL, NWO-I and Stichting Physica.

This book is published simultaneously in Dutch: H. Brookhuis, *ASML en de Nederlandse natuurkunde: Een geschiedenis van het Advanced Research Center for Nanolithography (ARCNL, 2014-2024)* (Amsterdam: Amsterdam University Press, 2025).

This English book is translated from Dutch by Erin Goedhart-Stallings.

Cover illustration: Floris Krelage
Cover design: Petra Klerkx
Lay-out: Crius Group, Hulshout

ISBN   978 90 4857 163 5
e-ISBN   978 90 4857 164 2 (pdf)
e-ISBN   978 90 4857 329 5 (accessible ePub)
DOI   10.5117/9789048571635
NUR   697

Creative Commons License CC-BY NC ND
(http://creativecommons.org/licenses/by-nc-nd/4.0)

© H. Brookhuis / Amsterdam University Press B.V., Amsterdam 2025

Some rights reserved. Without limiting the rights under copyright reserved above, any part of this book may be reproduced, stored in or introduced into a retrieval system, or transmitted, in any form or by any means (electronic, mechanical, photocopying, recording or otherwise).

Every effort has been made to obtain permission to use all copyrighted illustrations reproduced in this book. Nonetheless, whosoever believes to have rights to this material is advised to contact the publisher.

Table of Contents

Foreword – Wim van der Zande   7

Foreword – Martin van den Brink   9

1. An experiment in itself   11
2. The competition   31
3. Bell Labs on the Amstel   65
4. Towards a new ARCNL   99
5. Conclusion   127

ARCNL at a glance   139

Acknowledgements   141

Archives consulted   143

Bibliography   145

Foreword

Why is the history of a scientific research institute being written after only ten years? The Advanced Research Center for Nanolithography (ARCNL) is unique in that it was the first public-private research institute in the Netherlands in which a company (ASML) established a long-term partnership with the academic world. ARCNL is a research institute with a mission. ASML's initiative was motivated by the belief that academic research would be more impactful if conducted by a cohesive team of researchers who share a common mission and long-term vision, a team that meets regularly in the coffee room.

The first few years at ARCNL were full of obstacles. ASML and ARCNL staff were required to be flexible in unfamiliar ways. It is those obstacles and the lessons learnt that inspired this book. It describes the evolution of ARCNL from a concept and a vision into a high-functioning organisation that operates at a top academic level while simultaneously reducing the time it takes to transfer ideas to ASML. ARCNL remains a dynamic research institute that grapples with many challenges, which makes it truly fascinating.

Wim van der Zande, Director of ARCNL

Foreword

When I started working at ASML in the early 1980s, innovation was largely driven by famous research laboratories like Philips Research (NatLab), IBM Research and Bell Labs, where I also came to know some of the leading researchers personally. Those laboratories employed small multidisciplinary teams of scientists and were responsible for groundbreaking technologies such as semiconductors, integrated circuits and the first lithography machines to produce those circuits. In partnership with companies, those laboratories achieved remarkable growth into the 1990s. Their strength lay in their small groups of researchers, which fostered collaboration across diverse competencies.

Much has changed since then. The famous research laboratories no longer exist in their old form. Products grew more complicated and that demanded innovations at the total product level, which is made up of many complex parts. Small teams have made way for large, complex organisations with new and diverse competencies. However, innovation in subareas remains essential. Around 2010, we at ASML realised that traditional laboratories like Philips Research could not tackle certain innovations. I envisioned the solution as integrating an ASML corporate laboratory into an open network of public research institutions, enabling deeper and broader research than any corporate lab could achieve alone. In the winter of 2012, I discussed this vision with Bart Noordam, which led to the idea for the Advanced Research Center for Nanolithography (ARCNL). The history of its development is described in this book.

The development of extreme ultraviolet (EUV) lithography highlights how innovative companies like ASML benefit from a robust knowledge infrastructure that stimulates their technological progress. EUV lithography is now used for the mass production of computer chips. This achievement is the result of 30 years of collaborative development, as no single institute had the financial and organisational resources to accomplish it alone. The success of EUV lithography stems from the intense collaboration between ASML and numerous research institutes, where each organisation contributed its own expertise to collectively develop a vision for the future.

It is important to prevent innovation research from becoming too focused on short-term goals. That is why research institutes must develop long-term goals and have adequate funding. The allure of such institutes attracts top researchers, who then cultivate new talent for research and ASML. ARCNL in Amsterdam, which is now celebrating its tenth anniversary, is a concrete result of this vision.

Integrating technological challenges into a research institute like ARCNL enhances the research and inspires the researchers. I believe the advantages of this far outweigh a theoretical approach rooted in complete academic freedom. Direction, provided at a distance via long-term agreements, gives researchers the space and security they need. At the same time, research institutes must continually reinvent themselves to stay relevant to society and innovative companies. Self-reflection and an open mind are essential in this regard.

Martin van den Brink, President & Chief Technology Officer at ASML until 2024

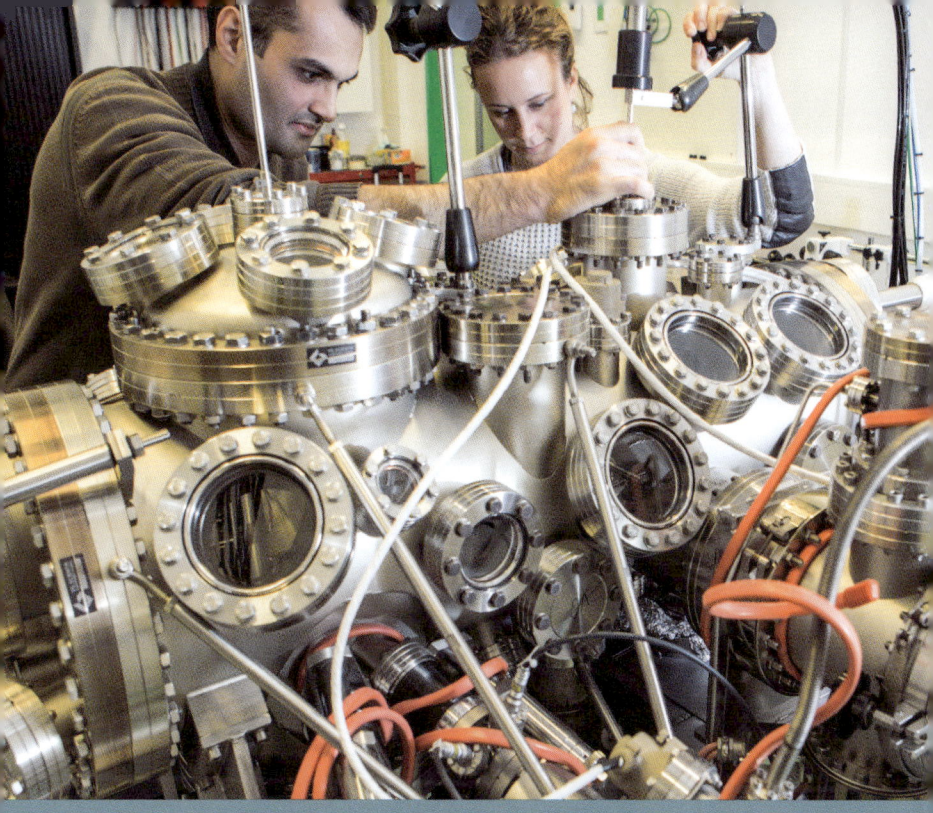

① An experiment in itself

The Advanced Research Center for Nanolithography (ARCNL) celebrated its 10th anniversary in 2024. ARCNL conducts basic research related to physics and chemistry, with a primary focus on the semiconductor industry in the Netherlands. Its research is especially focused on its partner ASML, the Veldhoven-based company that specialises in fabricating machines that produce microchips. Throughout its history there have been frequent references made to what stakeholders call the 'ARC model', and in 2017 ARCNL's management described the institute as 'an experiment in itself'.[1] This experiment is defined by the fact that the institute's mission is understood to combine scientific excellence with a research programme that is closely aligned with ASML's industrial interests.[2] In organisational terms, the institute is also an experiment in the Dutch scientific landscape: it is a close collaboration between ASML, the universities in Amsterdam, the Dutch funder of scientific research (NWO), and, for the past few years, the University of Groningen. The 10th anniversary is the ideal moment to analyse this experiment.

This book aims to identify and explain the various aspects of the ARCNL phenomenon, offering a unique look at the collaboration between science and industry in the Netherlands. The public-private character of ARCNL has attracted the interest of outsiders many times in its brief history. In a House of Representatives debate in April 2016, then Minister of Education, Culture and Science Sander Dekker called ARCNL an example of public-private collaboration and a driver of valorisation at public knowledge institutions.[3] The Rathenau Institute (in The Hague), which specialises in science policy issues, featured the fledgling ARCNL as a case study in their 2018 report, *Industry seeking university: The emergence of strategic public-private research partnerships.*[4]

That same year, the Dutch press wrote extensively about the then young institute. The sometimes difficult collaboration between ASML and the scientists in Amsterdam was sharply characterised in a September 2018 headline in the NRC *Handelsblad* newspaper: 'Blunt behaviour of chipmakers leads to tensions'.[5] In the summer of 2018, journalist Frank van Kolfschoten had asked within ARCNL how the collaboration was going. Director Joost Frenken had given his staff carte blanche to talk to the journalist: 'We have nothing to hide. There is nothing for us to be ashamed of.'[6] Furthermore, the then director was very conscious of the work-in-progress nature of ARCNL: 'We are in our early stages and together with our stakeholders [...] we are beginning to understand how to do all this.'[7] In 2021, Joost Frenken revealed himself to be an outspoken advocate of the public-private partnership model that had been achieved at ARCNL. Intensive collaboration with a company requires time and effort, but in his view, it is 'more than justified by valorisation, innovation and mutual inspiration.'[8]

Although the institutional collaboration between ASML and Dutch scientists was unprecedented in the Netherlands, it was part of a longer history of institutional innovation between the academic world and the semiconductor industry. In the 1970s and 1980s, consortia and 'university–industry research centers' emerged in the United States as new forms of institutional collaboration.[9] The ARCNL case study shows that the pursuit of new forms of cooperation between universities and the semiconductor industry also took place in the Netherlands, and it remains a highly relevant issue.

This book offers an analysis of what stakeholders have described as a learning process and experiment. First, I will focus on the considerations of stakeholders and those involved. Politicians, journalists and policy researchers have already offered their perspectives on ARCNL, but to what extent do those views reflect the trade-offs made in boardrooms and in scientific practice?

In addition, the chip sector is extremely politically charged and internationally competitive. How have scientific and business technology interests been balanced? And to what extent did they think in terms of conflicts or opportunities in that context? These questions are part of a broader reflection on the relationships between Dutch academic research and its applications, and on the relationships between Dutch natural scientists and the industrial world.

Second, the history of ARCNL will be set against the background of reforms in the Dutch scientific landscape. In 2010, Mark Rutte's first cabinet took office; it consisted of the liberal-conservative VVD and the Christian Democratic CDA with parliamentary support from the nationalist PVV. The top sectors policy was a key component in innovation policy and was proposed to stimulate cooperation between universities and companies. To this day, that policy also contributes funding to ARCNL's budget (via a TKI supplement from the Ministry of Economic Affairs).[10]

It is noteworthy that, as of 1 January 2017, the Foundation for Fundamental Research on Matter (FOM) officially ceased to exist. Until that point, FOM had organised physics research in the Netherlands, while the Dutch Research Council (NWO) represented other disciplines. After decades of unsuccessful attempts to do so, FOM became part of NWO. FOM had already had an official programme to facilitate public-private partnerships since 2004, including ones with Philips, Shell and ASML.[11] The FOM board was closely involved in the establishment of ARCNL. A history of ARCNL is therefore also a history of the institutional relationship between FOM and NWO, and the relationship between FOM and the Dutch industrial sector.

A final point to consider is the role of ASML itself and thus the role of the domestic and international semiconductor industry in the Dutch scientific community. Founded in 1984, ASML is now 40 years old and has experienced tremendous expansion over its history, with growing influence and geopolitical consequences. In 2023, ASML spent almost €4 billion on research and development (R&D) worldwide, about €400 million of which, in its own words, was spent on 'pure explorative research'.[12] By comparison, that same year, the Dutch government's total public R&D investment, including fiscal measures, was estimated at €7.5 billion.[13] The combined budget of Germany's 18 prestigious Helmholtz Institutes was around €6 billion in 2021. When the top sectors policy was launched, the 'High-Tech Systems and Materials' sector, which included ASML and Philips, accounted for almost half of private R&D investments in the Netherlands.[14] In addition to ARCNL, ASML also invests in research at other Dutch universities. For instance, Eindhoven University of Technology announced a large-scale investment by ASML on their campus in 2023.[15] The significant investments in semiconductor research are justified to maintain Moore's law, which states that the number of transistors on a microchip doubles every two years. Thus, a history of ARCNL also reveals how Moore's law has influenced Dutch science.

The above-mentioned considerations lie at the heart of this book. Its key question is how to interpret the establishment and functioning of ARCNL in terms of the social, cultural and scientific-technological contexts of the early 21st century. I shall start by explaining the position of this book on ARCNL in the wider Dutch history of science.

Between utility and curiosity: Public-private partnerships in Dutch science

'[T]he increasing familiarity that academic scientists, engineers, and medical researchers now have with industry, its concerns, its rhythms and routines, is alien to almost all humanists and social scientists.'[16] In his book *The Scientific Life* (2008), Steven Shapin, an American historian and sociologist of science, noted that many of his colleagues in faculties of social sciences and humanities still had little understanding of the scientific world at the beginning of the 21st century. When it came to the relationship between academic and industrial research, he observed, many of his colleagues fell back on binary assumptions that academic research should be 'free' and 'unfettered' and industrial research was merely 'applied' and conducted under 'control'. Shapin found such assumptions simplistic. Instead, modern natural science and engineering were characterised by 'social experiments' in which scientists investigated 'what novel configurations of people, space, knowledge, material resources, and external support can best bring about wanted intellectual and technological futures'.[17] In Shapin's view, by the end of the 20th century, we had entered the era of the 'entrepreneurial scientist' who actively seeks to build bridges between science and industry.

At first glance, ARCNL seems to fit nicely into Shapin's characterisation, in the sense that the institute presents itself as an innovative form of public-private partnership with a strong focus on basic research. Nonetheless, a history of ARCNL must also include some key nuances to pinpoint its place in the history of science. The way laboratories work, both at universities and in industry, has changed in the course of history.[18] But links between science and industry – and between research and application – have always existed. In this respect, the traditional ivory tower of academia may never have really existed in scientific research; at most, it may have been a rhetorical phenomenon.[19] Some 19th- and 20th-century scientists and their institutes can be characterised as 'entrepreneurial'.[20] Words like 'usefulness' and 'application' were also frequently heard at Dutch universities from the 19th to the 21st century. The interpretation of that usefulness shifted with the historical context. In the 19th century, the importance of usefulness was unsurprising, according to science historian Bert Theunissen. After all, the Dutch government funded universities primarily for their role as educators and lecturers. If the scholars of that era legitimised research, it had to be under the guise of usefulness or social relevance. An appeal to science as *'l'art pour l'art'* would fall on deaf ears.[21]

The 20th century marked the emergence of the professional researcher who needed to be prepared for a job in government or industry.[22] For example, one important future employer of Dutch students was Philips. In the 1930s, so many physics graduates from VU Amsterdam went to work at Philips that – in keeping with the university's religious identity – a special psalm was composed and sung whenever a student made the switch.[23] This relationship between university and industry meant that the application itself did not occur at the university, but that the university was the breeding ground for talent for companies like Shell and Philips.[24] In the 1930s, these companies advocated for the development of courses in applied physics, and they continued to influence university research after the Second World War.[25] A caveat here is that professors with an eye for industry were still in the minority. So while Dutch companies had contacts at universities in the 20th century, the situation in the other direction was still less developed.[26]

The research policy of Dutch universities is just one perspective on the importance of scientific research for industry. A lot of research into the history of science focuses – implicitly or explicitly – on academic scientific research and rarely on the research that is done in industrial settings.[27] However, the history of industrial laboratories is equally relevant to ARCNL as the history of academic laboratories. As later chapters will describe, ARCNL was inspired by a variety of sources: the AMOLF institute in Amsterdam was the academic inspiration, but the new organisation was also inspired by famous industrial laboratories like AT&T's Bell Labs and Philips' NatLab.

Those laboratories faced their own challenges. How does one organise an industrial laboratory? How much time should a researcher spend on basic research and how much should be spent on applied research, if such terms are used at all? Can researchers publish freely or not? Historian Ernst Homburg describes these considerations as a form of 'navigating by instinct'. This is less a characterisation of the research work itself and more about the constant factor in the organisation of the industrial research laboratory: 'Fundamental strategic choices also contain an arbitrary element. In this sense, large multinational corporations also navigate to a large extent by instinct.'[28] This book shows that the search for the ideal collaboration between ARCNL's partners was also a form of navigating by instinct.

A central question in this book is how different perceptions of science were negotiated between the partners involved. What is the ideal way to carry out public-private science and who should do it? What demands could be placed on what Shapin calls the 'entrepreneurial scientist'? This question is part of a broader reflection on the changing view of the scientist, which historians of science describe as scientific personae. The identity of scientists is not about the individual biographies of researchers; instead it is centred around the ideals and changing demands placed on scientists, which are partly shaped by institutes.[29] Institutes play a major role in the scientific reward system and thus help shape the identity of scientists. For example, securing research funding had become an 'essential skill' for American scientists by around 1940, and this would later also apply to European researchers.[30] The collaboration of stakeholders from both the academic world and research funding and industrial research sheds light on how scientists are being shaped in the 21st century.

NWO, FOM and the top sectors: An institutional sketch of Dutch science policy

Starting in 1945, Dutch physicists worked under the auspices of FOM, the Foundation for Fundamental Research on Matter. 'A big whale in the relatively small pond of ZWO' is how chemistry professor Hugo Kruyt characterised the relationship between FOM and ZWO (the predecessor of NWO) in the Dutch funding landscape for scientific research in 1953.[31] When FOM acted, it created a stir.[32] In 1959, another chemist, this time Hendrik Westenbrink of Utrecht University, described FOM as the 'cuckoo's egg' in ZWO's nest.[33] These were not declarations of love. The large budgets and far-reaching autonomy of Dutch physicists were not uncontroversial, but they nonetheless managed to maintain their privileged position until 2017. A key achievement of the FOM board in its final years was their close involvement in establishing ARCNL. With its three institutes – AMOLF (atomic & molecular physics), NIKHEF (subatomic physics) and DIFFER (energy, formerly plasma physics in Rijnhuizen) – FOM had succeeded in putting a firm stamp on the institutional landscape of Dutch science.

The history of FOM has attracted considerable interest among historians in recent years.[34] The self-perception of FOM that can be gleaned from existing research is that of a confident group of physicists with a well-developed political antenna and an ever-watchful eye for collaboration with industry. Although the founding and growth of the budgets for physics research were often seen as a continuation of pre-war science policy – as if they were a logical follow-up to the establishment of the Netherlands Organisation for Applied Scientific Research (TNO) in 1932 – it was the physicists themselves who played an important role in the post-war evolution. Hendrik Kramers and his colleagues used the post-war prestige of physics (linked to the development of the atomic bomb) to successfully argue that the government should spend more money on basic research.[35]

The boost for basic research lasted until the early 1970s. At that time, the government paid little attention to the governance of science. However, the Science Policy Advisory Council was set up with scientists and industry representatives in 1966, and the first minister of science policy was appointed in 1971.[36] Until then, the scientists themselves had determined how the available budgets would be spent. The 1974 Science Policy Memorandum is often considered a turning point in this regard: alignment with social priorities, both idealistic and economic, became the focal point of science policy.[37] In tough economic times, the usefulness of spending on scientific research was called into question.

The 1970s and 1980s were characterised by a seemingly paradoxical development. Spurred on by student protests and the installation of the progressive Den Uyl cabinet (1973–1977), scientists were asked to make a larger social impact in the 1970s. As the 1980s began, we saw a growing focus on market forces. The government also wanted a better understanding of and more control over spending on science, and ZWO had to be transformed and centralised.[38] Idealism was quickly replaced by financial pragmatism.[39] Historian Jorrit Smit also observed this change in the spatial design of universities: the science shops where citizens could go with their questions in the 1970s were replaced by transfer points and science parks that universities used to attract business investment.[40] As early as 1980, these places were cynically described as 'a kind of science shop for the not-underprivileged'.[41]

The history of Dutch science policy therefore should not only be understood in terms of *what* science should do or *how* it should be organised, but also *where* it should take place and *who* should carry it out. ARCNL is located right next to AMOLF at Amsterdam Science Park, and is also a spatial demonstration of a public-private partnership. How public-private partnerships in physics come about is thus also a question of the spatial embedding of the research. What did the ideal laboratory look like to those involved? To what extent did ARCNL resemble existing models and what needed to be reinvented? In this respect, this book is in line with recent research that treats the increasing focus on innovation in Dutch science policy as a question of *where* useful knowledge is developed and transferred.[42] This book about ARCNL is also an attempt to use the laboratory at Amsterdam Science Park to investigate how a new generation of natural scientists has dealt with the changing political and social expectations of science since 2010.

'The long arm of Moore's law': Science & the semiconductor industry

When discussing the historical role of companies in 20th-century R&D, Dutch historians often refer to the 'big five': Shell, Philips, AkzoNobel, DSM and Unilever. Their size and investments made them important to understanding the broader development of science in the 20th century. For example, Philips regularly collaborated with university scientists, so the company also appears in the history of uranium enrichment and Dutch space research.[43] These companies dominated private R&D spending in the Netherlands for a long time, but for the past few decades ASML has been making its presence known.[44] In 2003, ASML had about 1,500 people working in its R&D department.[45] By 2023, that number had grown to over 15,000 worldwide.[46] Between 2009 and 2019, ASML also surpassed its Brabant neighbour in terms of Dutch R&D spending. In 2009, Philips still spent over €270 million more than ASML on research in the Netherlands, but ten years later, the situation had clearly changed. In 2019, ASML spent €1.4 billion on research in the Netherlands, almost twice the amount Philips spent.[47]

Dutch machine manufacturer ASML has become much more prominent in the chip industry in recent years, a shift reflected in the media. In 2020, the BBC described ASML as 'a relatively obscure Dutch company', a quote proudly printed on jumpers for employees in Veldhoven.[48] Only four years later, all Dutch politicians were familiar with the company, and almost every international magazine or newspaper had written about it. Alongside ASML, similar technology companies have caught the eye of historians and journalists in recent years.[49] This book is not just another account of the history of ASML but a reflection on its role in Dutch science. Hybrid research centres like ARCNL have long been encouraged in the United States, and initiatives in microelectronics in particular have been seen as a model for such collaboration.[50] Historian Cyrus Mody argues that in the chip industry Moore's law should not be understood merely as a predictive law, but as an engine that drives and steers developments.[51] It has led to institutional innovation with the emergence of all kinds of hybrid research centres; therefore it was partly Moore's law that changed 'how American science was done'.[52]

Institutional innovation in the chip sector is inseparable from the global political and economic dynamics of the semiconductor industry. The fact that Moore's law changed American science had everything to do with the political importance that American politicians placed on the chip sector: in the 1980s, the industry grew to become a prime example of a perceived decline of US economic and industrial strength and the growth of Japan.[53] Nowadays there is talk of a chip war between the United States and China.[54] Towards the end of the Cold War, economic competitiveness became part of American national security: the large SEMATECH (Semiconductor Manufacturing Technology) consortium of the 1980s, which was launched to support the US chip sector, was partly funded by the US Department of Defense.[55] It was America's answer to Japan's Very Large-Scale Integrated Semiconductor Project (VLSI). Europe tried to offset this with a coordinated industrial R&D policy via the Eureka programme. It was partly these latter grants that financed part of ASML's R&D programme.[56] Building on Cyrus Mody's work, this book shows that high R&D costs affected both the industry involved and Dutch academic science.

Outline

The themes mentioned above form the structure of this book. The history of ARCNL will be discussed chronologically with each chapter focusing on a specific theme. The history of ARCNL revolves around science policy, the creation of institutes and the training of contemporary scientists. How did those involved perceive the ideal form of science that should be practised at ARCNL? And how did those perceptions of science and values translate into the expectations of the directors, principal investigators and PhD candidates? What determined which research was considered useful or not? This book will use these themes to show how universities, ASML and the scientists involved attracted and repelled one another as they worked to successfully complete the first ten years.

Chapter 2 will examine the origins of ARCNL, which arose from ASML's desire to establish a nanolithographic research institute. It covers the advent of the top sectors policy, the background to

ASML's interest in fundamental physics and how ARCNL ultimately emerged from a collaboration between the needs of FOM, AMOLF, ASML, the universities in Amsterdam and NWO.

Chapter 3 will cover the period starting in 2013, when the abstract and technical policy desires for public-private partnerships took concrete shape in the collaboration around ARCNL. The emphasis here is on the stakeholders' differing visions about how ARCNL should function and what could be expected of the scientists at the new institute. Initially conceived as a 'Bell Labs on the Amstel', the young institute was faced with the question of what it actually wanted to achieve and how it could go about doing so.

Chapter 4 will focus on the period that began in 2017, a time that has become known as 'ARCNL 2.0'. This period reflects the maturing of the institute, which also came to fruition in 2019 with the new Matrix VII building at Amsterdam Science Park. This chapter focuses mainly on the concrete reality of scientific life at the institute, which shifts between academic science and research for industry. Who are the entrepreneurial scientists and what is expected of them?

The history of ARCNL centres around several questions that are essential to understanding contemporary science and scientists. What does it mean to conduct research that is relevant to both academia and industry? How are the various interests weighed, and what does this mean for the demands placed on scientists? Finally, the conclusion will reflect on these questions.

Notes

1. ARCNL *Self-Evaluation 2014-2016*, June 2017, 3.
2. 'ARCNL Mission', https://ARCNL.nl/our-mission (Accessed 23 September 2024).
3. House of Representatives, 'Report of a Plenary Debate on 1 June 2016', 2015–2016 session, 33 009, no. 17.
4. Tjong Tjin Tai, S.Y. et al. *Bedrijf zoekt universiteit – De opkomst van strategische publiek-private partnerships in onderzoek* [*In-*

dustry seeking university: The emergence of strategic public-private partnerships in research] (The Hague: Rathenau Institute, 2018).
5. Frank van Kolfschoten, 'Bot gedrag van chipmachinemakers leidt tot spanningen' ['Blunt behaviour of chipmakers leads to tensions'], *NRC Handelsblad*, 6 September 2018.
6. Message from Joost Frenken to the ARCNL Governing Board, 12 July 2018. ARCNL Archive.
7. Message from Joost Frenken to ARCNL staff, 12 July 2018. ARCNL Archive.
8. Joost Frenken & Udo Kock, 'Publiek-privaat onderzoek kan prima autonoom zijn' ['Public-private research can be perfectly autonomous'], *Het Financieele Dagblad*, 27 October 2021.
9. Cyrus Mody, *The long arm of Moore's law: Microelectronics and American science* (MIT Press, 2016); Elizabeth Popp Berman, *Creating the Market University: How academic science became an economic engine* (Princeton University Press, 2011).
10. TKI: Top Consortium for Knowledge and Innovation. This was later called the PPP (public-private partnership) supplement.
11. For example, see Pieter de Witte, 'Public-private partnerships – An example from the Netherlands: The Industrial Partnership Programme', in: *Nanotechnology Commercialisation for Managers and Scientists* (2012), 263–290.
12. ASML *Annual Report 2023*, 21.
13. Includes spending on R&D, innovation and fiscal support. See: Broek-Honingh, N.G. et al. *Totale Investeringen in Wetenschap en Innovatie 2018-2024* [*Total Investments in Science and Innovation 2018-2024*], (The Hague: Rathenau Institute, 2020), 9.
14. J.P. van den Toren et al. *Coördinatie in de topsectoren. De geplande TKI's en hun uitdagingen* [*Coordination in the top sectors: The planned TKIs and their challenges*] (The Hague: Rathenau Institute, 2012), 12.
15. 'ASML and Eindhoven University of Technology strengthen longstanding collaboration', Eindhoven University of Technology press release, 24 April 2023, see: https://www.tue.nl/en/news-and-events/news-overview/24-04-2023-asml-and-eindhoven-university-of-technology-strengthen-longstanding-collaboration
16. Steven Shapin, *The Scientific Life: A Moral History of a Late Modern Vocation* (Chicago University Press, 2008), 265.
17. Ibid., 264.

18. Ernst Homburg, *Navigating by instinct: A historical view of industrial and university research* (Maastricht, 2003), 6.
19. Steven Shapin, 'The Ivory Tower: The history of a figure of speech and its cultural uses', *The British Journal for the History of Science* vol. 45, no. 1 (2012), 1–27.
20. R. Daniel Wadhwani, Gabriel Galvez-Behar, Joris Mercelis & Anna Guagnini, 'Academic entrepreneurship and institutional change in historical perspective', *Management & Organizational History* vol. 12, no. 3 (2017); Joris Mercelis, Gabriel Galvez-Behar & Anna Guagnini 'Commercializing science: Nineteenth- and twentieth-century academic scientists as consultants, patentees, and entrepreneurs', *History and Technology* vol. 33, no. 1 (2017).
21. Bert Theunissen, *'Nut en nog eens nut.' Wetenschapsbeelden van Nederlandse natuuronderzoekers 1800–1900* ['*Usefulness and more usefulness': Visions of science of Dutch natural scientists from 1800–1900*] (Hilversum: Verloren, 2000), 191.
22. Pim Huijen, 'Universiteit, bedrijfsleven en de opkomst van de beroepsonderzoeker 1880–1940' ['Universities, the business world and the rise of the professional researcher 1880–1940'], in *Onderzoek in opdracht- De publieke functie van het universitaire onderzoek in Nederland sedert 1876* [*Commissioned research: The public function of university research in the Netherlands since 1876*] (Hilversum: Verloren, 2007).
23. Ab Flipse, 'Geen weelde, maar een offer', Vrije Universiteit, achterban en de natuurwetenschappen' ['Not opulence, but a sacrifice: VU Amsterdam, constituency and the natural sciences'], *Universiteit, publiek en politiek. Het aanzien van de Nederlandse universiteiten* [*Universities, the public and politics. The reputation of Dutch universities*] (Hilversum: Verloren, 2012), 76.
24. David Baneke, 'Toegepaste natuurkunde aan de universiteit – contradictie of noodzaak?' ['Applied physics at university – contradiction or necessity?'], in *Universitaire vormingsidealen – de Nederlandse universiteiten sedert 1876* [*University educational ideals – Dutch universities since 1876*] (Hilversum: Verloren, 2006), 29–38.
25. Peter Baggen, Jasper Faber & Ernst Homburg, 'Opkomst van een kennismaatschappij' ['Rise of a knowledge society'], in: Johan Schot et al., *Techniek in Nederland in de Twintigste Eeuw, deel 7* [*Engineering in the Netherlands in the 20th century. Part 7*] (Zutphen: Walburg Pers, 2003), 140–173.

26. Homburg, *Speuren op de tast* [*Navigating by instinct*], 25.
27. Ann Johnson, 'What if we wrote the history of science from the perspective of applied science?' *Historical Studies in the Natural Sciences* vol. 38, no. 4 (2008), 610–620.
28. Homburg, *Speuren op de tast* [*Navigating by instinct*], 13.
29. Lorraine Daston & Otto Sibum, 'Introduction: Scientific personae and their histories', *Science in Context* vol. 16, no. 1/2 (2003), 1–8.
30. Robert Kohler, *Partners in science: Foundations and natural scientists, 1900–1945* (University of Chicago Press, 1991), 395.
31. Quoted from Albert Kersten, *Een organisatie van en voor onderzoekers: de Nederlandse organisatie voor Zuiver-Wetenschappelijk Onderzoek (Z.W.O.)* [*An organisation by and for researchers: the Netherlands Organisation for Pure Scientific Research (ZWO)*] *1947-1988* (Assen: Van Gorcum, 1996), 71.
32. Ibid.
33. Quoted from Kersten, *Een organisatie van en voor onderzoekers* [*An organisation by and for researchers*], 131.
34. Hoeneveld & Van Dongen, 'Out of a clear blue sky? FOM, the bomb, and the boost in Dutch physics funding after World War II', *Centaurus* vol. 55, no. 3 (2013), 264–293; Friso Hoeneveld, *Een vinger in de Amerikaanse pap: Fundamenteel fysisch en defensieonderzoek in Nederland tijdens de vroege Koude Oorlog* [*A finger in the American pie: Basic physics and defence research in the Netherlands during the early Cold War*] (PhD dissertation, Utrecht University, 2018); Abel Streefland, *Jaap Kistemaker en uraniumverrijking in Nederland 1945–1962* [*Jaap Kistemaker and uranium enrichment in the Netherlands 1945–1962*] (Amsterdam: Prometheus, 2017); Dirk van Delft et al., *Snaren, spiegels en plakband – 70 jaar Nederlandse natuurkunde* [*Strings, mirrors and sticky tape – 70 years of Dutch physics*] (W-Books, 2017).
35. Hoeneveld & Van Dongen, 'Out of a clear blue sky? FOM, the bomb, and the boost in Dutch physics funding after World War II.'
36. David Baneke, 'De vette jaren: de Commissie-Casimir en het Nederlandse wetenschapsbeleid 1957–1970' ['The good years: the Casimir Commission and Dutch science policy from 1957–1970'], *Studium* vol. 5, no. 2 (2012), 125.
37. David Baneke, 'De veranderende bestuurscultuur in wetenschap en universiteit in de jaren zeventig en tachtig' ['The changing governance culture in science and academia in the 1970s and

1980s'], *BMGN- Low Countries Historical Review* vol. 129, no. 1 (2014), 32.
38. Patricia Faasse, *Profiel van een faculteit: De Utrechtse bètawetenschappen 1815–2011* [*Profile of a faculty: Sciences at Utrecht University 1815–2011*] (Hilversum: Verloren, 2012), 154–155.
39. Baneke, 'De veranderende bestuurscultuur in wetenschap en universiteit in de jaren zeventig en tachtig' ['The changing governance culture in science and academia in the 1970s and 1980s'], 34; 52.
40. Jorrit Smit, 'Kennisoverdracht op de campus. Transferpunten, bedrijfscentra en science parks in de jaren tachtig' ['Knowledge transfer on campus. Transfer points, business centres and science parks in the 1980s'], in Ab Flips & Abel Streefland (Eds.) *De universitaire campus: Ruimtelijke transformaties van de Nederlandse universiteiten sedert 1945–2020* [*The university campus: Spatial transformations of Dutch universities from 1945–2020*] (Hilversum: Verloren, 2020), 119–143.
41. 'Plan universiteit: wetenschapswinkel voor regio' ['University plan: science shop for the region'], *Nieuwsblad van het Noorden*, 17 June 1980. Quoted from: Jorrit Smit, 'Kennisoverdracht op de campus' ['Knowledge transfer on campus'], 124, note 18.
42. Jorrit Smit, *Utility spots: Science policy, knowledge transfer and the politics of proximity* (PhD dissertation, Leiden University, 2021).
43. Streefland, *Jaap Kistemaker*; David Baneke, 'Organizing space: Dutch space science between astronomy, industry, and the government', in: Thomas Heinze & Richard Munch (Eds.), *Innovation in science and organizational renewal: Historical and sociological perspectives* (New York: Palgrave Macmillan, 2016), 183–209.
44. Homburg, *Speuren op de tast* [*Navigating by instinct*], 49.
45. Ibid; ASML, *Annual Report 2003*, F-41.
46. ASML, *Annual Report 2023*, 8.
47. 'Top 30 Bedrijfs-R&D in Nederland 2010' ['Top 30 R&D Companies in the Netherlands in 2010'], *Technisch Weekblad*, 17 April 2010; 'R&D 2020 Top 30', *Technisch Weekblad*. Documents obtained from: https://www.rankingthebrands.com/The-Brand-Rankings.aspx?rankingID=300&nav=category (Accessed 9 April 2024).
48. Marc Hijink, *Focus: De wereld van ASML. Het machtsspel om de meest complexe machine op aarde* [*Focus: The ASML Way. Inside*

the power struggle over the most complex machine on earth] (Amsterdam: Balans, 2023), 119.

49. Jorijn van Duijn, *Fortunes of high-tech: A history of innovation at ASM International, 1958–2008* (Techwatch Books, 2019); René Raaijmakers, *ASML's architects: The story of the engineers who shaped the world's most powerful chip machines* (Techwatch Books, 2018); Paul van Gerven & René Raaijmakers, *NatLab – Kraamkamer van ASML, NXP en de CD* [*NatLab: Breeding ground for ASML, NXP and the CD*] (Techwatch Books, 2016); René Raaijmakers, *De geldmachine – De turbulente jeugd van ASML* [*The cash machine – The turbulent youth of ASML*] (Techwatch Books, 2017).
50. Mody, *The long arm of Moore's law: Microelectronics and American science*, 224.
51. Ibid., 8.
52. Ibid., 16.
53. Mario Daniels & John Krige, *Knowledge regulation and national security in postwar America* (Chicago University Press, 2022), 209–210.
54. Chris Miller, *Chip war: The fight for the world's most critical technology* (Simon & Schuster, 2022).
55. Larry Browning & Judy Shetler, *SEMATECH: Saving the U.S. semiconductor industry* (Texas A&M University Press, 2000); Daniels & Krige, *Knowledge regulation*, 194–230.
56. Hijink, *Focus*. For an overview of early ASML R&D budgets, see: Rutger Bregman & Jesse Frederik, 'Maak kennis met de grootste uitvinder aller tijden' ['Meet the greatest inventor of all time'], *De Correspondent*, 25 February 2015.

② The competition

'As committed spectators, we wish you a fair competition with creative and decisive play.'¹ On 16 February 2013, Hendrik van Vuren, head of research policy at FOM, announced the start of a competition organised by ASML. Four locations in the Netherlands and Germany competed: the universities in Nijmegen, Eindhoven and Aachen and a consortium from Amsterdam made up of AMOLF, VU Amsterdam and the University of Amsterdam. At stake in the competition was nothing less than an entirely new research institute: the Institute for Nanolithography (INL). It was initiated by ASML, a manufacturer of semiconductor machines. Hendrik van Vuren played a coordinating role in this procedure on behalf of FOM. The announcement of the competition marked the end of ten days of deliberations between all parties involved, especially FOM, which would be expected to provide financial support to the new research institute. The result of the competition was announced in May 2013: the Institute for Nanolithography would be located in Amsterdam.²

What we now know as ARCNL was quickly built from the ground up, 'in a flurry of activity' as the period was described later.³ In less than a year, ARCNL was a reality. NWO, FOM, AMOLF and the universities were involved in addition to ASML, and local politicians also played a role in the bid procedure. Professors were surveyed, finances had to be arranged, and a location had to be selected. To understand how these diverse parties came together in the blink of an eye for the ASML initiative, I will start this chapter by contextualising the institute's place in discussions about science policy. Then I will analyse the dynamics in the ARCNL pressure cooker.

Although competition was not a new phenomenon in Dutch science, this particular competition was. Scientists were used to submitting grant applications to research funders such as NWO, FOM and Technology Foundation STW. But in this case, ASML had taken the lead by proposing a new institute and evaluating the candidates' funding proposals. That was unusual because research funding was normally distributed and organised through NWO. So how should this competition be organised? What motivated ASML to launch this initiative? Why did it seek to establish closer ties with universities in 2013? And how did the academic community respond to a competition launched by ASML? The entire procedure was a swift and intense collaboration between the funding organisation (NWO), the academic community and Dutch industry.

To begin, I will outline the state of Dutch science policy in 2013. Of particular importance is the top sectors policy: the plan of the Rutte I & II cabinets to encourage companies and universities to collaborate more. How was this plan received in science and industry, and what role did it play in the creation of ARCNL? I will also examine ASML's motivation. What were the technical challenges and why did ASML seek to establish closer ties with universities in 2013? This approach will give us a better understanding of the motivations and the individuals who played a role in creating ARCNL.

From innovation paradox to top sectors policy

'We are going to do things fundamentally differently than all previous cabinets', announced Maxime Verhagen, then Minister of Economic Affairs, Agriculture and Innovation, in September 2011.[4] Although the comment was intended as a joke – given Verhagen's previous cabinet experience – the top sectors policy marked a clear break from the past. The new principle was that companies would not necessarily receive targeted subsidies for R&D but would instead receive tax deductions for innovation activities.

More important for the science sector was the announcement that existing research budgets would be directed towards these 'top sectors'. At the time, these were nine sectors in which the government, industry and knowledge institutions had to work together. According to the Minister, 'that is where all research budgets should go'.[5] These sectors can be traced back to the 1980s, when they were still known as innovation-oriented research programmes. They were renamed Top Technology Institutes at the end of the 1990s. For a long time, these initiatives had used Dutch natural gas revenues to incentivise public-private partnerships. Starting in 2011, the Dutch Research Council (NWO) would set aside part of its budget for grants.

Dutch scientists had been keeping a wary eye on the top sectors policy for some time. The focus on innovation may have sounded promising to scientists, but the head of NWO, physicist Jos Engelen, was less than enthusiastic: 'This disappointing plan is taking money out of our own pocket'.[6] After all, the money now came from NWO, which was expected to redirect almost half of its scientific research budget to the top sectors.[7] For scientists, the whole plan felt like a financial step backwards: the new cabinet stopped allocating the FES funds, the pot of Groningen natural gas revenues that had been used in part to finance scientific research since the 1990s.[8]

Both before and during the top sectors plan, Dutch policymakers were captivated by what was known as the 'innovation paradox', a phenomenon that can be traced back to the 1980s, when it was known as the 'innovation gap'.[9] Although Dutch science performed exceptionally well, the industrial community benefited very little from it. In 2011, the Advisory Council for Science and Technology (AWT) called this the 'knowledge paradox'. Indeed, the Netherlands was 'one of the countries with the highest scientific output and citation scores' in Europe, but 'the use of this excellent knowledge' lagged behind.[10] It all amounted to the same thing: in the opinions of policymakers and policy advisers, something was going wrong in the transfer of scientific knowledge to the business world.

The Young Academy, a partnership of young researchers under the banner of the Royal Netherlands Academy of Arts and Sciences (KNAW), was outraged by the way in which the government tried to solve the so-called innovation paradox. NWO funds were being diverted from basic science to innovation-oriented research. In the opinion of the Young Academy, the glowing reports on Dutch science meant there was no reason to make fundamental changes.[11] Had not many reports shown that Dutch science was actually functioning very well?[12] So why did they have to forfeit money to the less well-performing innovation sector?[13] A historical review by the Rathenau Institute was similarly critical: had the government not learnt any lessons from all the policy interventions it had launched in the decades prior to the top sectors policy?[14] Those critical researchers saw the government's increasing interference in science primarily as symbolic politics from The Hague. Though it gave the impression of a decisive government, it offered no guarantee of success.

Neither the researchers at the Rathenau Institute nor those at KNAW were enthusiastic about top-down policy regarding innovation. In their view, innovation was an unpredictable process, and the creation of knowledge between science, the business community and the government – seen as the 'golden triangle' – only described a fraction of the total innovation process. The innovation process was also hampered by obstacles such as financing, organisational culture and market developments. Some researchers therefore felt it was unfair to view the productivity of scientific research as a simple calculation: 'Judging the research in the golden triangle by counting the number of innovations is tantamount to judging things over which the participants have little or no influence.'[15]

The Rathenau Institute felt that the notion of government guiding innovation was a typically European trend, and Dutch policy was not unique in this respect.[16] Furthermore, the 'knowledge paradox' was a widespread concept among other European governments, not just in the Netherlands.[17] European funding schemes that primarily aimed to support industrial R&D had been in place since the 1980s. But the new plan in 2011 seemed to have a more direct impact on the budgets for academic research. In 2012, the details of this new policy became more concrete: with the establishment of the Top Consortia for Knowledge and Innovation, the government's financial support for public-private partnerships became a reality.[18]

ASML, a 'relatively obscure Dutch company'

The Dutch government's aim to increase collaboration between universities and the business community came at an opportune moment for ASML. Before 2012, ASML had been a relatively obscure company. As a manufacturer of microchip production machines it operated alongside giants like Intel, TSMC and Samsung, yet the ASML brand remained virtually invisible in end products like smartphones and laptops. Nonetheless, it was already a respected name in the scientific community. Unlike many large companies, the chip machine manufacturer was deeply integrated into a closely-knit local network. Prior to 2012, ASML spent almost 90% of its R&D funds in the Netherlands, a stark contrast to companies such as Philips, Unilever and Shell.[19] ASML's spending on R&D increased sharply from 2012 onwards, partly in the Netherlands but also abroad. The reason for this change could not be linked directly to the new top sectors policy, which aimed to make R&D more appealing from a tax perspective. Instead, it was largely related to the technological challenges the Veldhoven-based company was facing.

ASML's increasing R&D spending should be understood in the context of the international chip industry. The dynamics in this industry are governed by Moore's law: the prediction from the 1960s that the number of transistors per chip would double every two years, meaning that chips would become increasingly smaller and more powerful, but also increasingly cheaper per transistor. That is worth a lot of research money. Moore's law should not be understood as a law of physics that merely describes or observes a reality, but as a social phenomenon. In a highly decentralised and globalised production chain, this law allows for coordination among the many companies and institutes that are part of the microchip production chain.

ASML specialises in lithography, equipment that uses light to print patterns on chips. This step determines the size of the transistors on the chip. Since the end of the 1980s, ASML had been investing in new lithography technology to uphold Moore's law. There were still several technological options on the table until the end of the 1990s, but starting in 2001 ASML set its sights on EUV machines as the most viable route towards ever smaller and better microchips.[20] EUV stands for 'extreme ultraviolet' and this technology applies patterns to microchips using light with a wavelength of 13.5 nanometres. Economic historian Chris Miller called this choice the end of the 'lithography wars', a time of uncertainty over which technology would win out.[21] Between 1999 and 2001, ASML had briefly invested in the scalpel joint venture, a technology based on electron beams.[22] Although it was said to be highly precise, there were concerns that the technology might be too slow and therefore unsuitable for mass production. Through European programmes, ASML was also involved in an alternative approach in Austria and had set up a European consortium around the EUV (EUCLIDES). From 1999, ASML was the only non-American company allowed to participate in the EUV Limited Liability Company, a consortium of the American chip sector and various national labs that had a budget of $250 million.[23] Participation by Asian companies was a very sensitive issue: since the 1980s, American government policy had been aimed specifically at winning the chip war against Japan. But the only requirement for ASML to join the consortium was that it had to make part of its investments in America.[24]

Geopolitical issues quickly came into play around EUV lithography. The US government had established a wide range of trade restrictions since the 1980s, largely to protect American industry from the emerging Japanese chip sector.[25] In the 1980s, the chip sector was increasingly seen as part of US national security, in both an economic and a military sense. Even without actual political intervention, the politicised context had a deterrent effect that even the American EUV consortium could not avoid.[26] Japanese company Nikon was initially a candidate to join the consortium alongside ASML, but withdrew after political resistance arose.[27]

The Committee on Foreign Investment in the United States (CFIUS) also sprang into action in 2000 and 2001 when ASML wanted to take over SVG (the Silicon Valley Group), which was a lithography company that was also part of the EUV consortium.[28] Resistance mainly came from the US Department of Defense, which considered an SVG subsidiary to be of military relevance. There was only sporadic resistance in relation to the commercial EUV technology; a small minority expressed objections to a potential monopoly on EUV technology by ASML.[29] In the end, the Americans relented after an extensive review by CFIUS and intervention from both the Dutch Minister of Economic Affairs and the Minister of Foreign Affairs.[30] It would appear that the relevance of future EUV technology for American national security was mainly assessed on strictly military grounds and not from an economic perspective.[31]

During the controversial takeover of SVG, ASML could count on the support of much of the American chip sector, particularly market leader Intel.[32] They felt that collaborating with a foreign company that could supply the semiconductor equipment was necessary to keep the American chip sector competitive. Between 1996 and 2002, ASML's market share in lithography machines grew from 20% to 55%, while it also developed a dominant position in EUV technology.[33] The ambitious plans for EUV development received even more support around 2012: major players in the chip sector (Intel, Samsung and TSMC) injected a combined €1.4 billion of capital into ASML to facilitate development of the new technology.[34] This was in addition to a $4 billion investment from Intel in the same year.[35]

In the years prior to the founding of ARCNL, it became clear that ASML was not on track with its original plans and was unable to get the EUV machines to work on the desired scale. Luc van den Hove, CEO of the Leuven-based research institute IMEC, was quoted in the scientific journal *Nature* in 2012 saying that 'the R&D challenges ASML is facing are enormous'.[36] ASML's machines are an enormous challenge for physicists and engineers. A chip now has 10 to 100 billion transistors, each of which is just a few dozen nanometres in size, and 20 nanometres is about one-thousandth of the thickness of a hair. Essentially, ASML's machines project light that uses a blueprint to print the pattern of the transistors on silicon. But behind this simple description lie many complex processes. Both the blueprint and the silicon layer move in the machine, and the machine must know their positions to within a nanometre during the process. The light is created by shooting a laser at droplets of tin, which are heated to 500,000 degrees Celsius. High-quality mirrors reflect this light without absorbing it, preventing any distortion of the image. All these processes involve complex physics: the generation of the light source, the effects on the materials and the precise measurements that are constantly needed to ensure the system functions optimally.

The EUV light source was a major concern at ASML. To stimulate development, in 2012 the company purchased the American company Cymer, the leading supplier of components for the new light source. In the space of 15 years, a great deal of money and political capital had been invested in EUV technology. The 2013 initiative to stimulate partnerships with universities can thus be understood in the context of ASML's technological challenges and ambitions in the international market.

With the introduction of the EUV machines, ASML's top management had expected to double its revenue, but the limited power of the light source was a major obstacle.[37] ASML had set its sights on a 1000-watt light source, but according to Jos Benschop, Senior Vice President of Research, it was a 'struggle to get to 100 watts [...]. We didn't really understand it properly either.'[38] Bart Noordam, Vice President of Research, also felt that this was a 'huge problem' at the time. Improving the EUV light source was an

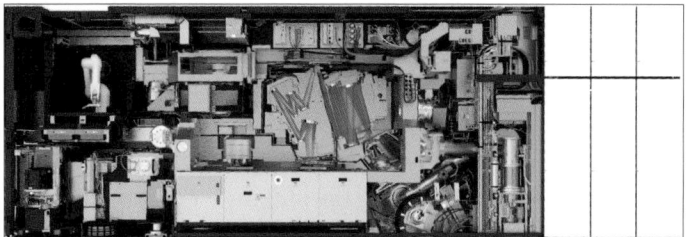

1. ASML's EUV machine: the basis of ARCNL's research programme. Photo: ASML.

economic necessity.[39] That is why it seemed to ASML that this topic was ideally suited for a new research institute. Not least because it could take ten years to finish it: 'We needed more scientific substantiation that could possibly lead to breakthroughs.'[40]

In 2012, ASML had hired Bart Noordam, a Vice President of Research who could foster relationships with universities.[41] Between 2009 and 2012, he had been dean of the Faculty of Science at the University of Amsterdam. Noordam had obtained his doctorate at the AMOLF institute, had been a member of the scientific staff there and had been the director of the institute from 2002 to 2005. He also had served as the director of development and engineering at ASML. Noordam could be seen as a bridge builder between the worlds of government, academia and business. In addition to his academic career, he had worked as a consultant at McKinsey and was the first director of the North Holland Auditor's Office from 2005 to 2008.[42] A colleague from his postdoctoral studies at the University of Virginia in 1992 described Noordam as a 'car salesman'.[43] In other words, Noordam can be seen as a scientist and director who moved easily between different worlds.

In the past, Noordam had regularly expressed his opinion on Dutch policy regarding young researchers, and he was also interested in partnerships with the industrial community. His commitment to the plight of PhD students was evident from his inaugural speech upon accepting the position of professor by special appointment of atomic and molecular physics in 2006, with the pertinent title: 'It is time to organise PhD education and research'.[44] That same year, he co-authored a handbook titled

Mastering your PhD, a survival guide for PhD candidates.[45] In addition, as dean he had already attempted to set up a substantial collaboration with industrial partners, mainly in the field of chemistry.[46] With his background in Dutch academia and his focus on the careers of young researchers, it seems no coincidence that with Noordam ASML took the initiative for ARCNL.

The plan

A first meeting between Noordam and Van Vuren quickly revealed the outlines of what ASML had in mind: a research centre focused on advanced nanolithography – especially extreme ultraviolet (EUV) technology – organised as a public-private institute. With an annual budget of around €10 million, the plan was to set up an institute of roughly 100 people and locate it at Eindhoven University of Technology, Radboud University in Nijmegen, AMOLF in Amsterdam or RWTH Aachen University.[47] The German university may sound like the odd one out, but Aachen had two key advantages: its campus hosts several prestigious Fraunhofer research institutes and it is home to research centres for corporate giants like Ericsson, Microsoft and Siemens. Moreover, the inclusion of Aachen balanced the ratio of technical and more general universities in the bidding process.

The ultimate aim was to enhance interaction between the scientific community and ASML through what was then called an 'open innovation programme'. The open innovation model reflects an approach in which innovative companies mainly share knowledge in a network and do not rely solely on knowledge acquired internally. ASML wanted to learn to 'breathe' more easily with the world of academic research.[48] A new institute with a €10 million budget would help ASML remain at the forefront of EUV research.[49] ASML would contribute €2.5 million annually to the core funding (50% of the total funding) and pay another €1 million annually to the projects through the FOM Industrial Partnership Programmes (IPP). Thus, ASML's contribution would be roughly one-third of the total budget. Initiator Bart Noordam envisaged a minimum duration of eight years with three important output targets: solutions that supported the ASML Technology Roadmap, publications and intellectual property, and trained researchers, with an estimated annual outflow of ten postdocs, six PhDs and ten master's students.[50]

The fact that Hendrik van Vuren, head of research policy at FOM, was actively involved in the endeavour had everything to do with the position of that organisation. FOM would be 'in the lead' on behalf of NWO and would promote the project: FOM wanted to become a cofounder of the Institute for Advanced Nanolithography.[51] Its role was particularly relevant for the Dutch bidders: FOM would create a level playing field for all the Dutch universities that ASML had already approached.[52] Moreover, FOM would be able to contribute about €1 million annually from its budget via the IPP to match ASML's contribution. FOM's involvement immediately made its ambitions clear: it was 'prepared to run the institute [...] as a FOM institute'.[53] The national role of FOM meant that the university in Aachen could not benefit from its involvement. The dean of the Faculty of Science at Radboud University in Nijmegen, Stan Gielen, later explicitly thanked FOM for its role, 'which enabled the Dutch institutions to make a good offer compared to RWTH Aachen University'.[54] In other words, FOM primarily served the interests of Dutch physicists to prevent the ASML initiative from being lost to a German university.

FOM's desire to be involved from the start was also driven by its strategic vision on the new top sectors policy. At the end of 2012, FOM was still struggling with what was called the 'top sectors dilemma': the engineering and natural science were expected to be particularly relevant to the top sectors, but this could result in a relative decrease in budgets for curiosity-driven research in these disciplines.[55] By helping to set up the institute, FOM gained a certain amount of control over the process. However, FOM attached several conditions to its involvement. First, the institute had to secure funding from NWO or the Dutch government, since FOM's IPP budget for 2013 had already been 'significantly oversubscribed'.

Second, the new institute had to meet the scientific quality requirements of NWO and FOM. Third, FOM wanted to remain 'actively involved' throughout the development and establishment of the new institute. Finally, the most important condition was that ASML would make a greater financial contribution than had been proposed to date. ASML wanted to contribute €1.6 million in cash and €900,000 in kind to the core funding. FOM would rather see the ASML share contributed entirely in cash.[56]

ASML's ambitions for the institute were concrete, perhaps a little too concrete according to some. The focus had to be on research that would lead to innovation for lithography equipment, preferably equipment that could be brought to market within five to ten years. ASML even had a specific research question regarding the EUV machine for the yet-to-be-founded institute: 'How can the output power of the source be improved to 1000 W at 13.5 nm or even at 6.7 nm?' Van Vuren seemed to think it was all a bit too focused: 'Doesn't ASML want to be surprised?'[57] FOM was concerned about attracting young, talented researchers: 'A research profile that appeals to the imagination is a "must"' to make the institute a successful scientific enterprise.[58] Too strong an emphasis on specific ASML challenges may not be the most effective strategy to attract highly skilled researchers.

The bid books

The project proposals or 'bid books' from the universities were intended to give an overview of the plans for the following areas: research programme, organisation, available facilities, budget and timeline. ASML said it was open to suggestions with regard to the research programme, but in light of Van Vuren's observation, the question of whether this attitude was truly welcoming remained unanswered. Ultimately, ASML's priority was the scientific manpower that the bidders could mobilise. The names of researchers, their CVs and their availability were presented as the most important element in the bidding procedure: 'ASML considers this as the most important question in the list [...].'⁵⁹ In other words, whoever could compile the best list of appealing researchers would likely win the institute.

Why was Amsterdam interested in this institute? One of the driving forces from Amsterdam was Albert Polman, then director of AMOLF. He joined forces with the University of Amsterdam and VU University Amsterdam, where the science faculties were then hoping to merge into an Amsterdam Faculty of Sciences. For AMOLF, the technological challenges posed by ASML touched on scientifically interesting topics related to lasers, plasma and nanophotonics. Moreover, an essential element of the Amsterdam proposal was the model for scientific research AMOLF embodied: a flat and efficient organisation in which new scientific themes and talents could be launched. As a 2011 evaluation noted, AMOLF excelled in identifying new research areas that were relevant to both universities and industry.[60] Contributing to the new ASML institute was therefore very much in line with the mission that AMOLF saw for itself in Dutch science.

In addition to the scientific motivations, the creation of a nanolithography institute would also strengthen the 'high-tech profile' of Amsterdam Science Park.[61] In terms of budget, AMOLF hoped to receive €500,000 per year from the municipality, while the new research groups would have to raise approximately €3.75 million per year in project funding to reach a total budget of €10 million.[62] In the final proposal, this was revised to €2.6 million, partly because there was now more certainty that the national government would provide a higher top sectors grant, but also because the municipality of Amsterdam had pledged a one-off €5 million grant for start-up costs.[63]

The bid book from Nijmegen was overseen by Stan Gielen, dean of the science faculty who later became chair of NWO. They also had tried to involve the regional government in this: they hoped that the province of Gelderland would make €3 million available. However, no agreement had been reached on that.[64] The Nijmegen proposal also assumed the researchers would be able to secure €1.6 million in annual external funding.

Eindhoven had the most ambitious plan in this respect: the researchers there would have to raise €3.5 million in annual external funding in the first five years, and €4 million annually in subsequent years.[65] Like the Nijmegen plan, the Eindhoven plan had received no guarantees from local politicians. There were ideas and suggestions, but no firm commitments.[66]

The result was not long in coming. On 25 April, the project leaders from all locations had a 'tough but fair discussion' at ASML with Martin van de Brink, Bart Noordam, Jos Benschop and Patrick de Jager. The next day, the group from Amsterdam heard the news: ASML would first continue talks with Amsterdam about the plan; talks with Eindhoven, Nijmegen and Aachen were put on hold.

From Veldhoven to the Randstad

'It's a brilliant stunt that we've poached this from Brabant's high-tech pond', Albert Polman exclaimed joyfully when he told his Amsterdam colleagues the good news.[67] The director of AMOLF had struck a nerve that was deeply felt in Eindhoven. Many had expected Eindhoven University of Technology to have the home field advantage, being just a stone's throw away from ASML's headquarters in Veldhoven. The competition between the four universities illustrates that location matters in science.

ASML was well aware of the feelings in Eindhoven. The communications department had prepared a special document for it because 'it is natural for people to wonder why this institute is not created in the heart of Brainport. The paranoid may even wonder if ASML is about to do as Philips and shift its center of gravity northward'.[68] To soothe feelings in Eindhoven, ASML emphasised how small the new institute actually would be compared to ASML's R&D activities in Brabant. Moreover, ASML emphasised that it wanted to focus on basic research and that 'extending a knowledge network is easier where there is still a large untapped potential'.[69] In that context, the already close ties between Eindhoven University of Technology and ASML were presented as a disadvantage rather than an advantage.

Instead of proximity, distance might have been good for the new research institute, especially since it was involved in basic research and technology exploration: 'A bit of distance to Veldhoven actually doesn't hurt.' Of all the candidates, Amsterdam was the farthest away from ASML, with a calculated travel time of 1 hour and 20 minutes. But ASML mainly presented this distance as an advantage: 'It will make it easier for the institute to preserve its academic freedom and to avoid getting sucked into ASML's day-to-day issues.'[70]

The primary reason for choosing Amsterdam seems to have been the prestige of AMOLF. ASML saw AMOLF as a 'world-renowned' institute that could hold its own internationally with Harvard, MIT and Caltech. Moreover, in terms of both management structure and culture, AMOLF served as an example of what ASML wanted to create with the new institute: a research institute based on a flat organisational structure with little interference and bureaucracy from external partners and universities and a great deal of mutual collaboration. The other bid books could not offer a comparable appealing model. Amsterdam Science Park also was seen as a 'dynamic environment' in which there were all kinds of opportunities to interact with other 'pioneering researchers'. The role of the Amsterdam universities seemed somewhat overshadowed in the choice to go with Science Park, although their suggestion for a master's programme in nanolithography seemed promising as a way to attract students to this field.[71]

The city of Amsterdam also seemed very enthusiastic, certainly in comparison with the competing cities. On 3 April 2013, the municipal council had already promised to support AMOLF with funding of up to €5 million.[72] In the eyes of local councillor Carolien Gehrels (social-democrat, PvdA), ARCNL would become an autonomous institute that would serve an important purpose for ASML: 'ARCNL will enable ASML to develop new knowledge that will help the company prolong Moore's law [...]'.[73] However, the city of Amsterdam also had its own vested interests: the institute would create employment, attract more students and contribute to the 'high-tech profile' of the Amsterdam region. Thus, local politicians saw it as an important boost for their Science Park. ASML's indirect presence in Amsterdam could act as a magnet for other companies. The city also hoped that ARCNL would generate spin-offs, enabling young entrepreneurs to set up new companies around ARCNL.[74] Thus local interests also played a role in the establishment of ARCNL.

NWO, FOM and the Institute for Nanolithography

While AMOLF and FOM knew on 26 April 2013 that ASML essentially had chosen Amsterdam, this had to be kept secret from outsiders for another month – although that proved virtually impossible in the small physics community. A noteworthy detail was the reason for the secrecy: the boards of both FOM and the Dutch Research Council (NWO) had yet to make firm financial commitments. Did NWO want to contribute funds to an institute that it had not set up itself? There were also big question marks about the legitimacy of the entire initiative. Were ASML's technological challenges fundamental enough to justify embedding it in Dutch scientific institutes for the long term?

According to FOM, the financial commitment from NWO was crucial: 'FOM itself has not two pennies to rub together'.[75] Without the extra contribution, estimated at €4 million per year, it would have been impossible for the institute to get up and running. 'You can't get something for nothing', said Hendrik van Vuren.[76] What's more, there was no guarantee that ASML's financial support would be structural. What would happen to the employees of the future institute if ASML pulled the plug after ten years? So some safety nets had to be put in place as part of the FOM policy in case the institute ultimately had to continue without the support of ASML.

Yet halfway through 2013, there was still great uncertainty about the top sectors policy and its effects on scientific research funding. It was clear to FOM how NWO was positioning itself:

according to FOM, NWO would primarily use any additional funds to be 'more effective and decisive in its efforts in the top sectors'.[77] The new cabinet expected NWO to contribute €275 million of its annual budget to the top sectors policy, of which €100 million would be spent on specific public-private partnerships. This had implications for FOM and other disciplinary boards: the money that could be spent via them would decrease at the expense of the top sectors. However, there was now an opportunity to set up new programmes via the top sectors.

Halfway through 2013, FOM was mainly positioning itself to 'respond as strategically as possible to the combined effect of cutbacks, extra resources and shifts towards top sector resources'.[78] The FOM board believed that the top sectors also offered 'significant opportunities to recoup investments' by bringing in extra research funding.[79] So the ASML initiative came at a very opportune moment. FOM said it was 'ready to go' and help NWO develop the new policy. FOM proposed that more than €40 million of the extra NWO budget be spent on them. FOM director Wim van Saarloos immediately put forward the collaboration with ASML as a relevant project because 'business needs to be done here before the summer' and FOM could not afford these costs itself.[80] FOM had quickly succeeded in getting the ASML initiative for a public-private institute on the NWO agenda.[81]

The issue of funding the new institute also touched on the organisational embedding of the initiative. There was resistance at both FOM and NWO, and not everyone was on the same page internally. Some FOM board members felt that ASML had managed to take the initiative for itself rather easily, and that AMOLF had gone along with it too willingly. It was suggested that 'AMOLF has accepted de facto the conditions set by ASML'.[82] The critical response from FOM came as a surprise to the organisers at AMOLF: 'We thought they would be over the moon and congratulate us. But not at all; they were actually a bit indignant that we had done this.'[83]

It was not a given that the FOM board would support the founding of the new institute. Setting up a research institute in such close collaboration with a company raised questions about the relationship between basic research and the interests of ASML, and how that balance should be reflected in the institutional design. Was ASML's contribution of 30% of the total budget not a bit small? This criticism was an outgrowth of the institute's supposedly narrow scientific focus, which was mainly in line with ASML themes. FOM already had existing industrial-partnership programmes for such initiatives. Did nanolithography fit into the scientific portfolio that FOM had in mind?

But these critical comments were balanced by the opportunities the initiative offered. Some saw it as a 'unique opportunity' to fund long-term research in times of budget cuts. The political importance of ASML in The Hague could also have a positive effect: FOM was aware that the image of the country as a good place to do business ('Netherlands Inc.') could be 'worth billions' if it helped ASML remain competitive internationally.[84] With the creation of a new public-private institute, FOM could serve as a 'role model' in the new top sectors policy.[85]

Not only were arguments used to convince the FOM board, but also a certain degree of trust. Bart Noordam, the initiator from ASML, was invited to sit in on a FOM board meeting to 'create broad support' for the new institute.[86] In this way, Noordam was able to directly explain to FOM how ASML saw the initiative. The board seemed to gain the impression that ASML would give the new institute considerable freedom: 'ASML does not want any influence at the project level'. In FOM's view, Noordam had emphasised that it would 'not be an ASML institute'.[87] With his background as former director of AMOLF, Noordam was able to win the trust of the FOM board on the sensitive issue of academic freedom. According to FOM chair Niek Lopes Cardozo, Noordam had 'great confidence in the institute's ability to manage itself'. FOM was therefore convinced of ASML's commitment to giving the future institute director 'free rein' in the research programme.[88]

The legitimacy of the new institute remained a point of contention at FOM. The board noted that as a public-private partnership, the proposed institute would represent a 'powerful implementation of the top sectors policy'. However, it was also collectively emphasised that although the proposed scientific programme for the envisaged Institute for Nanolithography was fundamental in nature, it was also 'very strongly driven by ASML's current business case' and that this was 'insufficient reason [to] establish a new fully-fledged institute'.[89] An NWO or FOM institute had to be based on a basic research programme that was primarily driven by scientific developments, not by 'an acute business problem' at ASML. This view was also supported by board members with an industrial background.[90] Therefore, FOM saw it more as a long-term industrial-partnership programme. Consequently, it was not always represented as a new institute, but rather as a collaborative venture that was primarily supported by AMOLF and FOM. The model they had in mind would resemble the High Field Magnet Laboratory in Nijmegen: a collaborative venture involving many partners, but not a formal FOM or NWO institute.[91]

The tensions at FOM and NWO surrounding the establishment of the new institute had consequences. The name was described with words like 'centre' and 'partnership', and explicitly no longer as an institute. The envisaged Institute for Nanolithography should not be called an institute because it would 'not last forever'.[92] FOM director Wim van Saarloos had explicitly stressed that 'the word "institute" [...] cannot be used'.[93] This was not only due to doubts about the institute's scientific basis, but also the relationship between FOM and NWO. NWO also had 'reservations' about the establishment of a fully-fledged ninth NWO institute.[94] Therefore, NWO had expressed a preference for the name 'Center for Nanolithography'. Van Vuren urged all parties involved to accommodate NWO so that 'the institute can quickly stand on its own two feet and de facto function as a fully-fledged FOM institute'.[95]

At the end of June 2013, an agreement was finally reached on the new institute, but NWO still had a very critical message for FOM regarding the entire procedure. The chair of FOM's executive board, Niek Lopes Cardozo, told his fellow board members that 'we must reflect on future requests of this nature, because we cannot allow our agenda to be filled on a first come, first serve[d] basis'.[96] ASML and AMOLF had managed to set up a new institute through a combination of circumstances, but FOM could not simply leave the initiative for its future scientific programme to companies.

The Advanced Research Center for Nanolithography

Public-private research raises many questions about the desired balance between both sides. Although organisational issues may seem unimportant at first glance, they ultimately determine scientific practice. For example, how are voting rights on the board allocated? This is a key point of discussion, especially in a public-private partnership. At first, the FOM board was rather surprised that the proposed voting rights on the institute's board would be 50-50. After all, with NWO, the University of Amsterdam and VU Amsterdam, more public than private parties were involved, and public funding also accounted for the majority of the funding. Why was this not reflected in the distribution of voting rights? FOM chair Lopes Cardozo saw this as a strength: he expected that, in the long run, there was a good chance that there would be more private than public parties involved. In his view, fixing the voting ratio at 50-50 would better guarantee the interests of the public parties in the long term.[97]

While FOM and NWO were trying to set up the financing and structure, the founders were busy negotiating with ASML. After the public announcement, they worked hard to draft a partnership agreement. But the stakeholders' ambitions did not always align: 'There are many parties involved and they have many desires and requirements that do not always coincide'.[98] For example, FOM wanted to stipulate that the initiative was open to other industrial partners. For its part, ASML wanted voting rights on the board to reflect relative contributions to basic funding. The universities took issue with this because they planned to make a relatively limited in-kind contribution via their own scientific staff.⁹⁹ Furthermore, ASML was not enthusiastic about the possibility of research groups continuing to grow based on external funding: in the long term, this would make the basic funding too small and would not benefit the stability of the institute.

Finding a balance between public and private interests was also reflected in the positioning and formulation of what the institute would actually be. FOM director Wim van Saarloos ultimately formulated a new model.¹⁰⁰ ARCNL could not be a regular institute; instead, it had to combine 'the strength and rights of an NWO/FOM research institute' with 'collaboration and shared responsibility, funding and governance with external partners'.¹⁰¹ The structure had to do justice to the distinct identities of ASML and the public partners.¹⁰² The complexity of the matter was apparent from the doubts people had about how best to communicate this to NWO: should they emphasise how unique ARCNL was or how well it fit in with existing institutes? At first glance, institutes like AMOLF and NIKHEF appeared to be quite different: NIKHEF (subatomic physics) was a partnership between universities, while AMOLF was not. That said, in Van Saarloos's opinion, the partnership at NIKHEF was only at the programme level, while the governance of the institute as such fell entirely under the responsibility of FOM. This would not be the case for ARCNL.

ARCNL was unique in that its governance structure would include many partners, which also had to be reflected in its name.¹⁰³ Wim van Saarloos proposed the concept of an 'Advanced Research Center', mainly to address NWO's concerns. The idea was to develop an 'attractive new model' for institutional collaboration between companies and university partners.¹⁰⁴ This idea broke the deadlock in the summer of 2013. In early 2014, NWO even complimented FOM on its conceptual ingenuity for 'coming up with the advanced research center concept'.¹⁰⁵ The discussion about the status of ARCNL was thus important at NWO and FOM. Yet it also led to irritation at ASML, which had hoped to set up ARCNL as

2. Joost Frenken in the PiMu building, where the first laboratories were built. Photo: Henk-Jan Boluijt.

a regular institute, with the additional objective of increasing its appeal to researchers. These discussions meant that ARCNL went through several names and acronyms before August 2013: it began as the Institute for Nanolithography and temporarily became the Center for Nanolithography.[106] Eventually these ideas were combined into the concept of an advanced research centre, and the Advanced Research Center for Nanolithography was born.

ARCNL's first director

The position of director was essential to the establishment of ARCNL. In Joost Frenken, who came from Leiden University, they had found someone with an 'excellent scientific track record' coupled with 'a keen eye for possible applications'. Frenken also had a great deal of experience in collaborating with industrial partners, an important requirement for this position.[107] He had secured a prestigious European Research Council (ERC) Advanced Grant in 2010, won the FOM Valorisation Prize in 2012, had founded two start-up companies, had experience in managing large consortia and had a strong academic reputation. In addition, he was familiar with the institutional culture at AMOLF, where he had obtained his doctorate and been a group leader. AMOLF director Vinod Subramaniam, Albert Polman's successor, also believed that Frenken was very aware of the 'exemplary role' that ARCNL would have to play in Dutch science.[108]

Although Frenken seemed the ideal candidate on paper, he was not the first to be considered. It proved challenging to find a suitable candidate for a completely new institute. Names that came up included Huib Bakker, one of the people behind the Amsterdam bid book on behalf of AMOLF. Marc Vrakking, director of the Max Born Institute in Berlin, was also mentioned. The board was considering more than 15 names from across the entire Dutch physics community. Although it was not explicitly stated, almost all candidates were Dutch or had at least obtained their doctorate in the Netherlands; there was always a link with FOM.

In fact, this was a headhunting exercise in which four qualities held sway. First and foremost, they were looking for a 'top scientist' whose research touched on nanolithographic themes. Personal qualities were at least as important. The future director had to be an 'excellent' communicator, an 'inspirational leader' and an 'organisational wizard'.[109] Interacting with the many partners, each with their own interests, would be an important task. In addition, the director had to be able to attract young 'top scientists'.[110] Understandably, some candidates had a research background with ties to ASML. That was sometimes an advantage, but in other cases a disadvantage, since they would first have to find support at ASML. With Frenken, ARCNL finally recruited a candidate who was on their A list; the others had mostly declined the honour.[111]

Conclusion

Although ARCNL may have been set up quickly, 'in a flurry of activity', that does not mean it came out of nowhere. Around 2013, university researchers were unsure about the future of basic research funding in the Netherlands. Would there still be enough funding available in times of austerity? Politicians were encouraging NWO to use more of its budget for partnerships with companies. In contrast, ASML had no financial problems and was enjoying an upward trajectory, but it was eagerly seeking more scientific input to better understand the challenges associated with the new EUV machines. AMOLF embraced the ASML initiative as a unique opportunity that dovetailed nicely with their mission: to launch new scientific themes and talents. The success of the ARCNL initiative was the product of these developments.

By early November 2013, all sides had finally reached an agreement: a memorandum of understanding had been signed by NWO, FOM, the University of Amsterdam, VU Amsterdam and ASML. A director also had been appointed, although the minute details of the partnership agreement were yet to be worked out. Nevertheless, the exploratory phase could begin. Joost Frenken got the ball rolling with a thought exercise about additional external (industrial) partners. What was ASML's true position on this? Frenken intended to immediately 'provoke' his new colleagues. He thought this was necessary to find out 'how ASML really feels' about these issues. 'That clarity is lacking at the moment.'[112] In the next chapter, we will see how his exploratory efforts fared.

Notes

1. Announcement from Hendrik van Vuren on behalf of FOM to representatives of AMOLF, RU, TU/E AND NWO, 16 February 2013. ARCNL archive.
2. 'Amsterdam wint strijd om instituut ASML' ['Amsterdam wins battle for ASML institute'], NRC Handelsblad, 27 May 2013.
3. Letter from Van Vuren to ARCNL staff, 'Progress on Blueprint for ARCNL 2.0', 29 November 2017. ARCNL archive.
4. Robert Giebels, 'Nederland gaat fundamenteel anders van kennis naar kassa' ['The Netherlands has a fundamentally different approach to how knowledge translates into cash'], de Volkskrant, 14 September 2011.

5. Ibid.
6. Robert Giebels, 'Industriebeleid is verslechtering' ['Industrial policy is deteriorating'], *de Volkskrant*, 8 February 2011.
7. Ibid.
8. Evert-Jan Velzing, *Innovatiepolitiek: een reconstructie van het innovatiebeleid van het ministerie van Economische Zaken van 1976 tot en met 2010* [*Innovation politics: A reconstruction of the innovation policy of the Ministry of Economic Affairs from 1976 to 2010*] (Delft: Eburon, 2013), 119; 155–160; 167–174.
9. Jorrit Smit, 'Kennisoverdracht op de campus' ['Knowledge transfer on campus'], 130; Smit, *Utility spots*, 199. Also see: Ben Dankbaar, 'Omgaan met de innovatieparadox. Bestaat er een kloof tussen universiteiten en bedrijven?' ['Dealing with the innovation paradox. Is there a gap between universities and businesses?'], *M&O: Tijdschrift voor Management en Organisatie* vol. 59, no. 1 (2005), 64–80.
10. AWT, *Scherp aan de wind! Handvat voor een Europese strategie voor Nederlandse (top)sectoren* [*Full steam ahead! A guide to a European strategy for Dutch (top) sectors*] (2011), 23.
11. Beatrice de Graaf, Appy Sluis & Peter-Paul Verbeek, 'Weer moet er geld naar die topsectoren in de wetenschap' ['Yet again, money is going to those top sectors in science'], *NRC Handelsblad*, 14 June 2011.
12. Bart Funnekotter, 'Nederland is geen klein land in wetenschap; Haags beleid om onderzoek van universiteiten te sturen faalt, zegt het Rathenau Instituut' ['The Netherlands is not a small country in terms of science: The Hague's policy of directing university research is failing, says the Rathenau Institute'], *NRC Handelsblad*, 18 March 2011. Also see: Peter van den Besselaar & Edwin Horlings, *Focus en massa in het wetenschappelijk onderzoek: de Nederlandse onderzoeksportfolio in internationaal perspectief* [*Focus and mass in scientific research: the Dutch research portfolio from an international perspective*] (The Hague: Rathenau Institute, 2010).
13. Also see: Anna Grebenchtchikova, 'Topsectorenbeleid slecht voor kenniseconomie' ['Top sectors policy is bad for the knowledge economy'], *de Volkskrant*, 17 September 2012.
14. Harry Lintsen & Evert-Jan Velzing, *Onderzoekscoördinatie in de gouden driehoek. Een geschiedenis* [*Coordinating research in the golden triangle: A history*] (The Hague: Rathenau Institute, 2012).

15. Ibid., 37.
16. Stef Severt, 'Bedrijfsleven, niet overheid, moet innovatie stimuleren' ['Business, not the government, needs to stimulate innovation'], *Het Financieele Dagblad*, 26 October 2012.
17. Chiara Franzoni & Francesco Lissoni, 'Academic entrepreneurs: critical issues and lessons for Europe', in: Varga Attila (Ed.), *Universities, knowledge transfer and regional development: Geography, entrepreneurship and policy* (Cheltenham: Edward Elgar, 2009), 165.
18. W. de Haas et al. *Gouden Driehoek?; Discoursanalyse van het topsectorenbeleid [Golden triangle? A discourse analysis of the top sectors policy]* (Wageningen, Alterra University & Research Center, 2014), 43.
19. Jasper Deuten. *R&D goes global: Policy implications for the Netherlands as a knowledge region in a global perspective* (The Hague: Rathenau Institute, 2015), 35–36.
20. Vadime.Y. Banine. EUV *Lithography: Historical perspective and road ahead* (Eindhoven University of Technology, inaugural lecture, 2014), 12. For the evolution of technology options within the industry, also see: Dao et al., 'NGL process and the role of International SEMATECH', *Proceedings Volume 4688, Emerging Lithographic Technologies VI* (2002), 29–35.
21. Miller, *Chip war*, 183–185.
22. David Lammers, 'EUV gains as venture ends e-beam litho work', *EE Times*, 5 January 2001.
23. Miller, *Chip war*, 186–188.
24. David Lammers, 'U.S. gives OK to ASML on EUV effort', *EE Times*, 24 February 1999.
25. Daniels & Krige, *Knowledge regulation*, 193–230.
26. Ibid., 228.
27. William Holstein, 'U.S.-funded technology stays here, for now', *U.S. News & World Report* vol. 124, no. 19 (18 May 1998), 5; Linden, Mowery, Ziedonis, 'National technology policy in global markets – Developing next-generation lithography in the semiconductor industry', in: Maryann Feldman & Albert Link (Eds.), *Innovation policy in the knowledge-based economy* (Springer, 2001), 327.
28. Hijink, *Focus*, 79–80; 'ASM Lithography Holding NV to acquire Silicon Valley Group Inc. in an all stock transaction valued at EUR 1.8 billion (US$1.6 billion)', ASML *press release*, 2 October 2000; 'ASM Lithography N.V. completes acquisition of Silicon Valley Group Inc.', ASML *press release*, 22 May 2001.

29. Mark LaPedus, 'ASML refutes claim that SVG purchase will threaten EUV technology', *EE Times,* 17 April 2001.
30. 'ASML mag van Amerikanen SVG onder voorwaarden overnemen' ['Americans allow ASML to take over SVG under certain conditions'], *de Volkskrant,* 4 May 2001.
31. Miller, *Chip war,* 189.
32. Ibid., 188–189; Hijink, *Focus,* 79–80.
33. Linden, Mowery, Ziedonis, 'National technology policy in global markets', 326; 'ASML surpasses Nikon in '02 litho market', *EE Times,* 4 October 2003.
34. 'ASML issues shares to TSMC in connection with Customer Co-Investment Program', *ASML press release,* 31 October 2012.
35. Ian King & Cornelius Rahn, 'Intel investing $4.1 billion in ASML to speed production', *Bloomberg,* 10 July 2012.
36. Katherina Bourzac, 'A giant bid to etch tiny circuits', *Nature* vol. 487, no. 7408 (July 2012), 419.
37. Peter Wennink, 'Creating value for all stakeholders', *presentation at ASML Investor Day*, 23 November 2014. Accessed at: https://www.sec.gov/Archives/edgar/data/937966/000119312514423554/d826396dex992.htm
38. Interview with Jos Benschop, 4 June 2024, Veldhoven.
39. Interview with Bart Noordam, 26 June 2024, online.
40. Interview with Jos Benschop, 4 June 2024, Veldhoven.
41. In July 2012, it was announced that Noordam would move to ASML later that year. 'Personalia', *Het Financieele Dagblad,* 18 July 2012.
42. See: Reinier Bijman et al., 'Handelaar in bedrijfsgevoelige informatie' ['Trader in commercially sensitive information'], *De Groene Amsterdammer* vol. 47, 23 November 2011.
43. Dimitrios Papaioannou, *Multiphoton ionization of barium and sodium by intense picosecond laser radiation* (PhD dissertation, University of Viriginia, 1992), i.
44. Bart Noordam, *Het is tijd om promotieonderwijs en promotieonderzoek te organiseren* [*It is time to organise PhD education and research*] (Vossiuspers: Amsterdam, 2006).
45. Noordam & Gosling, *Mastering your PhD: Survival and success in the doctoral years and beyond* (Springer, 2022 [2006]).
46. Interview with Bart Noordam, 26 June 2024, online.
47. Hendrik van Vuren, notes of conversation with Bart Noordam, 18 January 2013. ARCNL archive.

48. FOM board, 'Draft: summary of conclusions and agreements from the Executive Board meeting' (no. 181), 14 May 2013. ARCNL archive.
49. Bart Noordam, 'First sketch Institute for Nano Lithography (INL) – Confidential FOM version', 24 January 2012. ARCNL archive.
50. Ibid.
51. 'Positioning of NWO and FOM re. ASML initiative for a public/private *Institute for Advanced Nanolithography*', Draft 6 February 2013. ARCNL archive.
52. Ibid.
53. Ibid.
54. Message from Stan Gielen to FOM, 10 April 2013. ARCNL archive.
55. Presentation to the Physics Administration Meeting, 14 December 2012. NA, NWO Archive, 2.25.109, 2336.
56. 'Positioning of NWO and FOM re. ASML initiative for a public/private *Institute for Nanolithography*', 16 February 2013. ARCNL archive.
57. Van Vuren's notes on document: Patrick de Jager, *Guidelines for Institute for Nano-Lithography bid book*, 12 February 2013. ARCNL archive.
58. 'Positioning of NWO and FOM re. ASML initiative for a public/private *Institute for Advanced Nanolithography*', Draft 6 February 2013. ARCNL archive.
59. Patrick de Jager, *Guidelines for Institute for Nano-Lithography bid book*, 12 February 2013. ARCNL archive.
60. NWO, *Evaluation 2005–2010: FOM Institute for Atomic and Molecular Physics (AMOLF)* (The Hague, 2011).
61. Memo from Albert Polman to AMOLF Policy Advisory Committee, 'Re: Institute for Nanolithography', 11 March 2013. ARCNL archive.
62. Draft texts for bid book ASML Institute for Nanolithography, 11 March 2013. ARCNL archive.
63. AMOLF, *Institute for Nano-Lithography (INL) – Strategic Plan/Bid Book*, 3 April 2013. ARCNL archive.
64. *Creating a leading Institute for Nano-Lithography*, Appendix 3. 10 April 2013. ARCNL archive.
65. Eindhoven University of Technology, *Institute for Nano-Lithography*, April 2013. ARCNL archive.
66. Harrie Verrijt, 'Regio zette laag in op ASML -plan' ['Region placed low bid for ASML plan'], *Eindhovens Dagblad*, 29 May 2013.

67. Message from Polman to AMOLF stakeholders, 'INL is coming to Amsterdam!', 26 April 2013. ARCNL archive.
68. ASML memo, 'Positioning the INL for regional media and stakeholders', 21 May 2013. ARCNL archive.
69. Ibid.
70. Ibid.
71. Ibid.
72. Gehrels to the Committee for Economic Affairs, Companies, Personnel & Organisation, 'Voortgang in het opzetten van het Advanced Research Center for Nanolithography (ARCNL)' ['Progress in establishing the Advanced Research Center for Nanolithography (ARCNL)'], 3 February 2014.
73. Ibid.
74. Ibid.
75. Hendrik van Vuren to the FOM Executive Board, 'ASML initiative for the founding of a new Institute for Nanolithography: results and follow-up bidding procedure', 7 May 2013, ARCNL archive.
76. Ibid.
77. FOM board, 'summary of conclusions and agreements from the Executive Board meeting' (no. 181), Held on Tuesday 14 May 2013 in Utrecht. ARCNL archive.
78. Ibid.
79. Ibid.
80. Wim van Saarloos, 'Subject: Response to your 2014 Framework Letter', 5 April 2013. NA, NWO archive, 2.25.109, 1691.
81. 'Towards a new PPP framework for NWO', Appendix to NWO Chairpersons' Meeting, 17 April 2013. NA, NWO Archive, 2.25.109, 1691.
82. Notes from Hendrik van Vuren, 'GVN/UB/RvB reports', 14 May 2013. ARCNL archive.
83. Interview with Huib Bakker, 25 June, Amsterdam.
84. Notes from meeting with FOM, 14 May 2013 (Confidential). ARCNL archive.
85. Draft, 'summary of conclusions and agreements from the Executive Board meeting' (no. 181), Held on Tuesday 14 May 2013 in Utrecht. ARCNL archive.
86. Message from Hendrik van Vuren to AMOLF, FOM, 'AMOLF Plan et al. for ASML initiative INL: our conversation on 13 May', 8 May 2013. ARCNL archive.
87. FOM Draft, 'summary of conclusions and agreements from the Executive Board meeting' (no. 181), Held on Tuesday 14 May 2013 in Utrecht. ARCNL archive.

88. Message from Niek Lopes Cardozo to Dirk Smit (and FOM board), 'Re: tomorrow after programme day extra Executive Board discussion on INL', 21 May 2013. ARCNL archive.
89. Hendrik van Vuren and Wim van Saarloos, Memo 'proposed Executive Board decision on INL', 16 June 2013. ARCNL archive.
90. Message from Dirk Smit to FOM board, 'Re: tomorrow after programme day extra Executive Board discussion on INL', 20 May 2013. ARCNL archive.
91. Draft BoD decision, 'Establishment of ASML-NWO/FOM-UvA/VU centre/collaborative venture in nanolithography', 20 June 2013. AMOLF archive; Notes of Hendrik van Vuren, 'Extra Executive Board meeting', 14 June 2013, ARCNL archive.
92. Notes of Hendrik van Vuren, 'Extra Executive Board meeting', 14 June 2013. ARCNL archive.
93. Draft BoD decision, Establishment of ASML-NWO/FOM-UvA/VU centre/collaborative venture in nanolithography', 20 June 2013. AMOLF archive.
94. Hendrik van Vuren to FOM Executive Board, 'Documents for your (additional) meeting of 14 June 2013; establishment of ASML/FOM Institute of Nanolithography', 13 June 2013. ARCNL archive.
95. Ibid.
96. Message from Niek Lopes Cardozo to FOM Executive Board, 'CNL|AB NWO is in agreement, planning discussion in BoD meeting tonight', 26 June 2013. ARCNL archive.
97. Marcel Bartels to FOM Executive Board, 'Documents for your meeting on 19 November 2013; Advanced Research Center for Nanolithography: from MoU to partnership agreement', 12 November 2013. ARCNL archive.
98. 'Notes on draft MoU', appendix to 7 June 2013 agenda, Second meeting for the Institute for Nanolithography, ARCNL archive.
99. Message from Bart van Leijen to Bartels, Van Vuren, Saarloos, and Polman, ' 'Draft MoU for INL', 1 June 2013, ARCNL archive; Message from Bart van Leijen, 4 June 2013. ARCNL archive.
100. Wim van Saarloos, 'Comparing the status of the Advanced Research Center for Nanolithography and regular NWO/FOM institutes', 26 July 2013. ARCNL archive.
101. Message from Wim van Saarloos to Albert Polman, 'RE: memo Wim incl. changes', 19 August 2013. ARCNL archive.

102. FOM, 'Documents for your meeting of 26 June 2013; establishment of ASML-NWO/FOM-UvA/VU Centre/Partnership for NanoLithography', 20 June 2013. AMOLF archive.
103. Message from Wim van Saarloos to Albert Polman, 'RE: memo Wim incl. changes', 19 August 2013. ARCNL archive.
104. FOM board, 'Summary of conclusions and agreements from the Executive Board meeting' (no. 863), Held on Tuesday 20 August 2013 in Utrecht. AMOLF archive.
105. Ibid. Reference to New Year speech by NWO director Hans de Groene on 6 January in 'Statements from the Director of Physics, January 2014', appendix to meeting documents for the 125th GBN meeting, 14 January. NA, NWO Archive, 2.25.109, 2326.
106. 'Meeting of the CNL Governing Board *in formation*', Meeting no. 3, 19 August 2013. ARCNL archive.
107. Renée-Andrée Koornstra to FOM Executive Board, 'Appointment of director of Advanced Research Center for NanoLithography (ARCNL)', 11 November 2013. ARCNL archive.
108. Vinod Subramaniam to FOM Executive Board, 11 November 2013, ARCNL archive.
109. Governing Board, Centre for Nanolithography (*in formation*), appendix to agenda for meeting on 26 September 2013. ARCNL archive.
110. 'Director of Advanced Research Center for NanoLithography (ARCNL)', appendix to agenda for GB meeting, 10 October 2013. ARCNL archive.
111. 'Meeting Report of CNL Governing Board *in formation*', meeting no. 6, 10 October 2013. ARCNL archive.
112. Joost Frenken to Bart Noordam, 'Thoughts on external partnership scenarios', attachment to Governing Board Agenda, meeting on 21 November 2013. ARCNL archive.

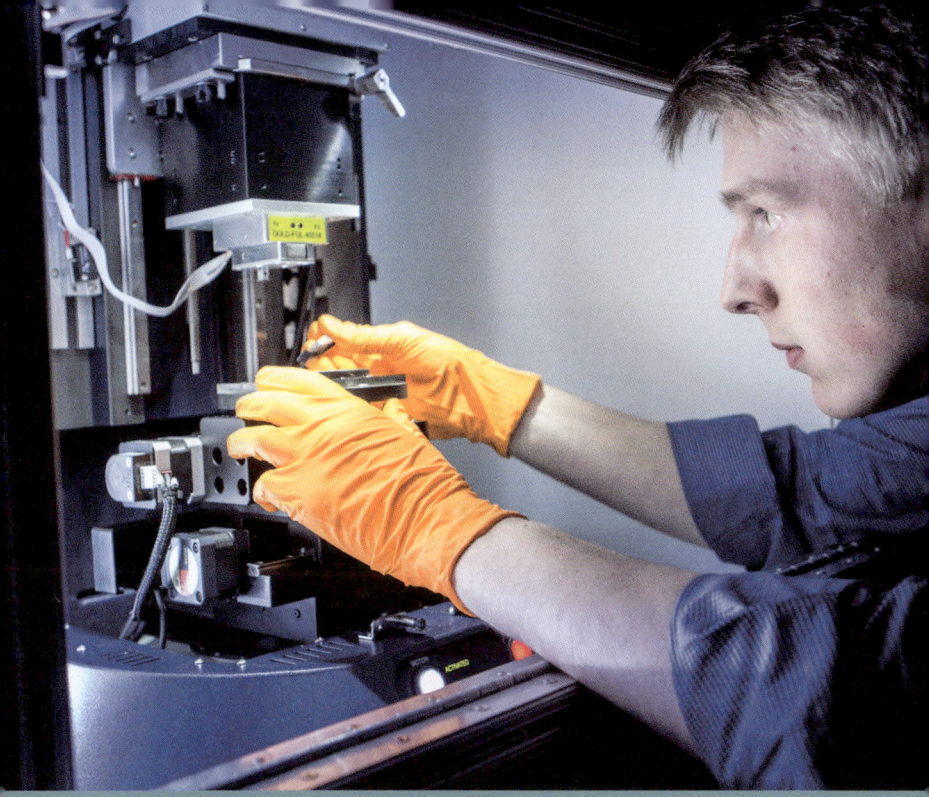

③ Bell Labs on the Amstel

'[W]hat is everyone's interpretation of "Bell Labs on the Amstel"?'¹ This question was discussed in the ARCNL board in May 2017, and it illustrates the complexity of the period from 2013 to 2017. Where did everyone stand in 2017? And what did the founders expect from the new institute? One might conclude that the high expectations had not been met. Perhaps more important was the realisation that it had not been entirely clear what the expectations had been in the first place. Was ARCNL meant to pioneer new technology? Was it supposed to be a leading centre of excellence in science? Was it intended to cultivate as much scientific talent as possible for the semiconductor industry? All of the above? Both the academic partners and ASML struggled with the question of exactly what purposes ARCNL was founded to serve and how it should be evaluated.

This chapter will examine the expectations and visions in the early years of ARCNL. Sociologists call these imaginaries: ideas, connotations, fantasies and convictions that play a role in how scientists shape their institutes. It is no coincidence that Bell Labs, in particular, emerged as a metaphor in discussions about ARCNL. Until the early 1990s, this industrial laboratory for the American Telephone & Telegraph Company (AT&T) was known as a paradise for basic research. Bell Labs has since grown into one of the most powerful symbols of a laboratory that facilitated autonomous basic research in an industrial context.[2] Bell Labs is sometimes called 'the most innovative scientific organization in the world', and to this day this notion tempts physicists to promote this former 'factory of ideas' as an ideal for contemporary innovation policy.[3]

Although in the Dutch context the Philips NatLab is often mentioned as an 'academic' industrial laboratory, Bell Labs was also part of Dutch physicists' world. FOM director Wim van Saarloos was active there as a researcher in the 1980s and 1990s, as were AMOLF director Albert Polman and ARCNL group leader Paul Planken. Bell Labs had become synonymous with the idea that commercial success and basic research could go hand in hand without the research being specifically focused on that end. Freedom and creativity, as well as interdisciplinarity and the long-term benefits of basic research, were all associated with Bell Labs.[4] It was precisely because of these associations that it continued to inspire hybrid research centres in which public and private parties worked together.

The success of Bell Labs is often attributed to specific historical factors, not least the fact that AT&T operated as a de facto monopoly and therefore had the financial means to maintain a lab with a great deal of freedom. It is precisely because of these factors that many have expressed doubts about the possibility of replicating this model. Nevertheless, the image of Bell Labs continues to capture the imagination. This chapter will use the dilemmas, compromises and synergies that were developed at ARCNL to explore how existing ideals of hybrid laboratories were translated to the contemporary Dutch context. This initially took place at AMOLF, where ARCNL was being founded, and later in its own temporary buildings erected right next to AMOLF. Instead of summarising the research results that were produced, this chapter will focus on the question of what research ARCNL was actually expected to conduct. Related to this is the question of what the partners hoped to achieve with the research at ARCNL.

In addition, the development of ARCNL took place within the broader evolution of Dutch science. After decades of failed attempts, FOM became a fully-fledged part of NWO and physicists lost their autonomous position in the scientific landscape. ARCNL was primarily established within FOM, and this chapter will therefore also reflect on how it tried to position itself as a national research centre in the context of institutional reforms.

Unlike Bell Labs, ARCNL was led by multiple partners from the academic world, industry and scientific funding organisations. Therefore, this chapter will focus on ARCNL's hybrid character as an institute. How autonomously could ARCNL operate? How did the partners' different visions of science relate to each other?

A false start

Although January 2014 is often seen as the official beginning of ARCNL, it was not yet a done deal at that point. The memorandum of understanding was little more than a declaration of intent, and the actual partnership agreement was still a long way off. Starting in the summer of 2013, many discussions were held about the concrete details of the planned research centre. By the end of 2013, Frenken had reached out to all the potential group leaders: 'Most of them are still on board, but they have critical questions about the freedom to make their own research choices at ARCNL'.[5] Lawyers were also consulted about whether the institute's proposed structure constituted a form of state support for ASML. Many questions were still unanswered, especially regarding ARCNL's autonomy from ASML.

The idealised image of ARCNL was that of basic research inspired by industrial issues at ASML, but the two worlds were not so easy to unite. In early 2014, for example, the fact that the difficult start was driving ASML's Senior Vice President of Research Jos Benschop 'round the bend' came up several times in conversations. In particular, discussions around intellectual property rights proved to be 'a thorny issue'.[6] The choice of Joost Frenken as director was a deliberate one, as he had experience in the industrial valorisation of academic research. But it was precisely that experience that caused a hitch. Frenken had previously founded a company called Applied Nanolayers (ANL), which worked with graphene. At the time, graphene was seen as a promising option for use in future microchips.[7] The company was based in Nijmegen, close to Dutch chip manufacturer NXP. Frenken himself had suggested that ANL's graphene research was also 'close to the world of ASML'.[8]

It was Frenken's entrepreneurial spirit that revealed a tension with ASML. Was it not advantageous that the director of ARCNL was working on a topic relevant to ASML, even if that work was with another company? Frenken thought this should be possible. In his view, it would be interesting to be able to collaborate scientifically with other, smaller companies, even if the subjects had a strategic overlap with ASML's interests. Frenken saw this as part of his scientific freedom.

However, Jos Benschop foresaw problems with this from ASML's perspective. By collaborating with ARCNL, those smaller companies could develop knowledge that could make them more valuable and therefore more difficult for ASML to acquire in the future.[9] Acquiring small companies is a proven strategy for gaining control of IP rights for strategic technology. The fact that Benschop and Van Saarloos would have to discuss this with each other on behalf of ASML and FOM was in fact already an escalation of the difference of opinion. 'If they cannot come to an agreement, there is reason to fear that ARCNL will not get off the ground', Hendrik van Vuren opined.[10]

Joost Frenken had founded Applied Nanolayers together with Paul Hedges (CEO) and Richard van Rijn, who had obtained his doctorate under Frenken and filled the role of CTO at ANL. For the ARCNL director this was a matter of principle: according to him, he would have had to leave a lot of research behind at Leiden University when he moved to ARCNL, and he wanted to prevent his graphene research from being left behind as well.[11] Moreover, he was also still formally affiliated with ANL: 'As co-owner of a company, I cannot be expected to engage in activities that would harm the interests of that company.'[12] The matter of principle that Frenken wanted to raise here concerned the question of whether research at ARCNL could also benefit companies besides ASML. In the end, the situation was smoothly resolved: ASML could claim potential IP rights, but ANL would be allowed to use them freely as long as they did not apply to ASML's own lithography technology. Frenken was satisfied: 'a real win-win situation (if I may use that cliché)'.[13]

During the start-up period, the expectations of ASML and ARCNL were not fully aligned. In March 2014, it became clear to Joost Frenken and Bart Noordam that there was a need for more intensive contact between ARCNL and ASML to prevent 'the perceptions and expectations at ASML and ARCNL from diverging'.[14] In addition to Frenken's entrepreneurship, the formation of the Scientific Advisory Council (SAC) was another hurdle that had to be overcome. The SAC would conduct an annual evaluation of ARCNL's scientific quality, but it was not self-evident to choose the people best suited to do that. Could scientific competitors sit on the council? Should people from ASML be included or not?

The original plan seemed to be based on the idea that both AMOLF and ASML would be represented on the council. Naturally, this immediately raised questions about the independence of the SAC. Joost Frenken felt that people with too close a relationship to ASML could not sit on the advisory council, 'because I can imagine that it will be very difficult for that person or others from ASML to look at the ARCNL programme from a purely scientific perspective without allowing ASML's technological interests to play a role in their assessment'.[15] Surely such a 'dual role for ASML' at ARCNL had to be avoided. ASML already exerted influence on ARCNL through its position on the board, and Frenken did not want the company to be able to influence the scientific quality evaluation as well. Some of the scientists that were put forward as candidates for the SAC also proved to be too 'sensitive a choice' for ASML.[16] Even without a representative from ASML, the company's involvement indirectly influenced the make-up of the SAC.

Wim van Saarloos from FOM argued that the make-up of the SAC offered opportunities. For instance, he campaigned for female scientists to be given a higher profile on the council: 'Let's make a good start in this respect too, right from the

3. ARCNL's first staff meeting. From left to right: Kjeld Eikema, Paul Planken, Fred Brouwer, Bart van Leijen, Joost Frenken, Wim Ubachs, Emile van der Drift, Joris van Bergen, Albert Polman, Ronnie Hoekstra, Stefan Witte. Photo: ARCNL.

beginning'.[17] He also thought it would be strategic to use the SAC to strengthen ties with Eindhoven and Nijmegen in order to involve the previous competitors in the bidding process in the new institute. In addition, he thought it would be good to be open to people from the business world, but from outside ASML.[18]

Van Vuren shared this opinion, but he thought that an 'ASML employee who can distance themself a little from the organisation should be up for discussion [...] or someone from Philips or TNO who does not receive funding from ASML'.[19] He also emphasised that the number of 'people from Amsterdam or with Amsterdam connections' in particular should be kept to a minimum, as ARCNL already had a very Amsterdam-centric composition.[20] After all, the group leaders primarily came from the University of Amsterdam and VU Amsterdam. Thus, the composition of the SAC was also an opportunity for ARCNL to not only be an Amsterdam-based ASML institute, but to forge stronger links with the rest of the Dutch scientific community.

Building a scientific programme: A 'man on the moon'

In scientific terms, the ARCNL programme was built around a core challenge set by ASML: 'a 1000-watt EUV source as a dream for the future'.[21] In more American terminology this was called a 'man on the moon' approach: an ambitious vision to guide the research projects.[22] In terms of autonomy, this meant that ASML set the broad framework and the scientists worked out the details. At AMOLF, coordinator Huib Bakker had converted the concrete problems and limitations of ASML's EUV machine into a scientific programme. For instance, in early 2013, the EUV source was still limited to 30–40 watts and the mirrors that reflected the EUV light had a very limited lifespan. AMOLF also noted that the photoresist, which the EUV light ultimately shines on, was not yet very efficient.[23] Investigating these problems at ASML also offered AMOLF opportunities to explore interesting scientific questions.[24]

A more concrete outline of the scientific programme was discussed in 2014. The programme was based around scientists from VU Amsterdam and the University of Amsterdam, and the group working on EUV Generation and Imaging was led by Stefan Witte and Kjeld Eikema. The EUV Targets programme reported to Paul Planken, and EUV Plasma Dynamics was set up by Wim Ubachs and Ronnie Hoekstra. After the first months it managed to recruit an average of one employee a week. By 1 January 2015, ARCNL had 37 employees.[25]

The issue of ARCNL's autonomy was already on the agenda during the first year. Stefan Witte, who had come from VU Amsterdam, had been awarded a prestigious ERC Starting Grant. ARCNL was also quite pleased with this, as it meant the intended investment target was quickly met. However, this success also immediately exposed a tension. Did that money rightly belong to ARCNL or to the university? And who was actually in charge of the group leaders? Was it the 'VU boss' or the 'ARCNL boss'? In early 2015, it was already being stressed that 'in each area, a new negotiation process with our friends at VU' was taking place.[26] The practical details of ARCNL were mainly being worked out as they went along.

Although the ARCNL board endorsed the 'AMOLF model' for the institute, allowing it to operate flexibly and autonomously, founder Bart Noordam felt that this was difficult to achieve, particularly due to the constant negotiations that had to be held with the University of Amsterdam and VU Amsterdam.[27] One cause of this situation was that the universities contributed to ARCNL in kind, mainly by seconding staff. ASML therefore believed that many of the problems would be solved if the universities simply started contributing in cash, even if this would de facto result in a smaller contribution from the universities.[28] This might give ARCNL more autonomy.

The initial tensions also reflected differing views on what constitutes good research. For instance, Bart Noordam and ASML were not big fans of what is known as revenue generation through applying for grants. This was a typical aspect of university research, but not of ASML's research. ASML did not need ARCNL to try 'desperately' to bring in €2.5 million in research funding every year, saying 'it is more important that the researchers are doing the research'.[29] But for the universities, being able to recover costs was a necessary condition for the functioning of ARCNL group leaders. The wide range of opinions meant that ARCNL's first challenge was to unite the worlds of the bid book, the partnership agreement and the partners' organisational cultures. Hendrik van Vuren, who would take on the role of chair of the ARCNL board, spoke in this context of 'running the gauntlet' to find a workable situation.[30] The idealised image of an autonomous and basic research institute where work would be done on ASML issues proved complex in practice. Tension had built both between ARCNL and ASML, and between ARCNL and the universities in Amsterdam.

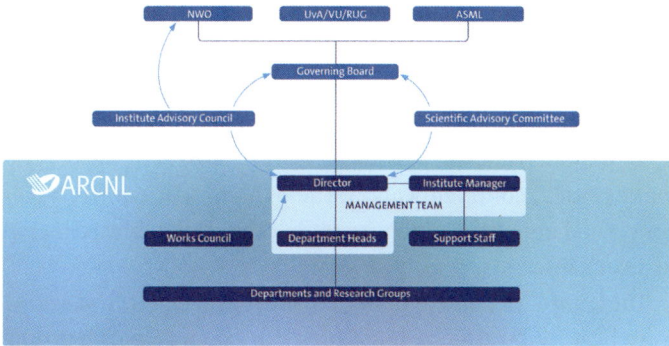

4. ARCNL's governance structure.

Between science and ASML

Despite the small initial hurdles, ARCNL's research received a largely positive response from scientific peers in early 2015. A panel of fellow scientists had evaluated the scientific programme and arrived at a relatively positive assessment. Frenken was jubilant: 'We've passed this litmus test! The overwhelmingly positive tone strengthens our confidence that we are on a meaningful path.'[31] The SAC was particularly pleased with the director himself. He was seen as a 'clear and major positive force' who, moreover, had managed to persuade good scientists to join in the ARCNL project and redirect their research towards nanolithography.[32]

The Dutch Research Council (NWO) also expressed positive feelings at the start of 2015. NWO director Hans de Groene said he saw ARCNL as an 'example' and an 'inspiration' for 'strategic partnerships with a longer than usual timeframe'.[33] In the perceptions of outsiders, a promising institute had been established in a short period of time. However, setting up scientific research that could appeal to both scientific peers and ASML seemed a complex issue.

At this early stage, ARCNL's scientific peers were extremely positive about its development. In their view, ARCNL was 'rapidly establishing itself as a nucleus for EUV-based research in the

Netherlands' and was very successful in combining research that was relevant to both ASML and the academic community.[34] The SAC viewed the eight delivered 'Invention Disclosures' – ideas for patents – as a strong showing. This council was mainly made up of people from the academic world. However, the SAC did make critical comments about the dynamics at ARCNL: it was concerned about the extent to which the research was geared to ASML's interests.

The complexity and dynamics of setting up research at ARCNL were best illustrated by Joost Frenken's own research group, which focused on materials research. His group was seen as an example 'of the way in which high-quality fundamental research [...] and industrial relevance can go hand in hand'. The SAC, however, saw changes to his research programme as being caused more by 'strategic concerns by ASML than concerns about the inherent scientific interest of the work'.[35] Frenken did not deny these dynamics, but mostly took a positive view of the changes. He saw opportunities for fundamental research *within* the technological challenges presented by ASML: 'a perfect ARCNL-style match!'[36] The scientific advisers seemed to be hoping for research that would be relevant to ASML but would still be primarily driven by scientific interests, and not the other way around. Perhaps they recognised the risk that ARCNL would focus primarily on relevance to ASML, relegating scientific relevance to a secondary role.

In 2015, the first tensions between the partners' differing visions of good science emerged: they assessed the progress of the research in different ways. Although fellow scientists were positive, ASML's judgement was more critical. One aspect of the public-private partnership was that ARCNL would produce Invention Disclosure Forms (IDFs). These were intended to identify ideas developed at ARCNL and communicate them to ASML for the generation of intellectual property (patents). Unlike the academic peers, ASML was dissatisfied with the number of ideas that were generated. At ASML, there was a feeling that the ARCNL scientists were too reluctant to share their ideas and too quick to pre-select ideas. Frank Schuurmans, Noordam's successor at ASML and on the ARCNL board, would 'rather see half an idea jotted down quickly than a fully developed idea that takes a long time to emerge'.[37] This represented a first sign of divergent cultures between ASML and the academic world.

After a difficult but enthusiastic start, ASML took an increasingly critical view of ARCNL's output. What's more, the evaluation of ARCNL was no longer in the hands of the people actively involved in its founding. This was worrying because ASML's concerns affected the future of the research institute. Frank Schuurmans even believed that based on the results at the end of 2015, ASML would not be inclined to continue its commitment to ARCNL. There was a perception that ARCNL had 'too little feel for ASML's problems'.[38] This was not a criticism of ARCNL's research as such, but rather of its lack of connection to ASML's interests. These views were widely shared at the company. ASML had chosen to act pre-emptively to send a message at an early stage. At the end of 2015, the first discussions were held to explore how ARCNL's impact on ASML could be increased, with a particular focus on 'how the research at ARCNL could be properly adjusted in a way that does justice to both the interests of academia and those of ASML'.[39]

ASML's side offered assurances that this was not a sign of 'retreat', and they still intended to 'make this joint ARC a success'. However, ASML wanted to make it clear that they were not entirely comfortable with the 'direction' and 'approach' of ARCNL.[40] ASML also received support from FOM in those discussions. By that time, Wim van Saarloos had been succeeded as FOM director by Christa Hooijer, and she too seemed to believe that ARCNL should focus on working more closely with ASML.[41] ARCNL group leaders could visit ASML once a month, and junior researchers could even be temporarily based in Veldhoven. One initiative that would have a long-term effect was the annual conference where staff from ASML and ARCNL came together to exchange knowledge.[42] The friction between them was mainly understood to be an 'issue of communication',[43] so the solution was also mainly understood to be a need to start a 'filtering procedure'.

To ARCNL management, it seemed that there was more to it than just a communication problem: it was a sign that the partners' expectations were not sufficiently clear. They were 'shocked' by the message from ASML and wanted clarity.[44] That led ARCNL to ask for a list of 'success factors'. If ASML was dissatisfied, Frenken wanted to understand their concrete expectations better. To this end, key performance indicators were drawn up 'to provide direction on how ARCNL operates and performs from ASML's perspective'.[45] For example, one indicator was the number of IDFs and patents produced, while another was the number of ARCNL employees who might be hired by ASML after completing their research. So it was not only the transfer of knowledge, but also the movement of workers that would play a role in the evaluation of ARCNL. There was also a conscious reflection on 'concepts and/or ideas that find their way into an ASML product', although it soon became clear that this would only become apparent after ten to fifteen years. In addition, publications with co-authors from ARCNL and ASML were seen as a positive indicator. The early friction between ARCNL and ASML was attributed to a lack of concrete expectations, and they hoped that quantifiable objectives would be the solution.

It soon became clear that they needed more than just quantifiable objectives. In April 2016, a special conference was organised to facilitate a more intensive exchange between ASML and ARCNL. ASML, in particular, spoke candidly about its wishes. After the conference, it was clear to ASML that 'the program of ARCNL requires alignment with that of ASML Research'.[46] This specifically involved coordination with ASML's research department, not with the company as a whole nor the Development & Engineering department. An integration board was set up to improve relations. Its members included Wim van der Zande, then an ASML director who had transferred from Radboud University in Nijmegen in 2014. In addition, the original bid book would be abandoned in the planning and evaluation of ARCNL. In their view, the institute should move more in the direction of ASML.

The room for manoeuvre at ARCNL lay in the tension between industrial relevance and academic independence. The young institute found itself in an awkward situation in which ASML stated that ARCNL was 'a logical and natural extension of ASML research', but Frenken countered that 'we hasten to add that ARCNL runs an independent research program'.[47] It was not easy to determine the desired degree of connection and distance. This was partly a conceptual discussion about applied and basic research in which some ARCNL researchers were concerned about 'a pressure to become more applied', while ASML feared 'a lack of impact'.[48] They also disagreed on how research should be organised. For example, ASML felt that research projects could be launched or ended more quickly, but Frenken felt that this 'should not be regarded [as] appropriate', thus defending his researchers. Moreover, he believed that ARCNL 'has demonstrated a very high level of compliance and susceptibility' with regard to ASML.[49] Hadn't ARCNL already moved enough in the direction of ASML?

The difference in approach between ARCNL and ASML had largely to do with the way in which research was structured. At ARCNL, research groups were set up according to academic convention: they consisted of a tenure track researcher (someone working towards a permanent position), postdocs and PhD candidates. PhD projects formed the foundation of the institute, and it was difficult to fundamentally change these four-year tracks simply because ASML had changed its mind, whether positively or negatively. If ASML was very enthusiastic, it preferred to transfer the project to ASML, where a whole team of researchers could get to work on it. If ASML was not enthusiastic, it preferred that the project be completely revised. Neither option was possible in the context of a PhD track at ARCNL.

The different perspectives on research were not merely a policy issue; they also affected the scientific staff. Some of ARCNL's staff increasingly felt that ASML was being given too much say. At least nine projects had been adjusted to a greater or lesser extent to align them more closely with ASML's interests.[50] In late

October 2016, it was announced that one staff member would be leaving for the Swedish Research Council. Allegedly, this was mainly due to dissatisfaction with the partnership with ASML and the way the company tried to influence the research projects. The person in question was Niklas Ottosson, a tenure tracker who had been hired in June 2015 and whose research group acted as a link between AMOLF and ARCNL, mainly on topics related to photochemistry. The main reason given for his departure was 'the way of interacting with ASML' and the 'impact this had on the atmosphere at ARCNL'.[51] ASML's communication was described as 'very direct' and therefore 'demotivating'. The board faced a challenge as this sentiment seemed to be widespread among ARCNL staff and was starting to set the tone at the institute. In Ottosson's view, ASML should adopt a 'hands-off' approach once a research focus had been determined.[52] ARCNL should not only have to defer to ASML, but ASML also had to learn to keep its distance from the research.

It was an uncomfortable situation, not least because Ottosson had been 'one of the promising talents' at ARCNL, a fact also acknowledged by ASML. However, Frenken recognised the signs: dissatisfaction with ASML's communication style was a widely shared frustration. People at ASML did not see what could be done about it: 'The company culture is simply that people are direct in their communication'. Moreover, Frank Schuurmans admitted to the board that 'many ASML staff [mainly] think that ARCNL should do what they want them to do'.[53] Neither the culture at ASML nor the way they communicated with ARCNL would be easy to change.

Thus, in the early years, the relationship between ASML and ARCNL was mainly characterised by different perspectives on research, in terms of design, implementation and communication. Yet this was not simply a conflict between academic science and industrial research. It was mainly a conflict between the visions that those involved had at the beginning, which were gradually changing to adapt to an unruly reality.

Between science and universities

As the first part of this chapter illustrated, it would be too simplistic to attribute ARCNL's growing pains solely to its relationship with ASML. The relationship with the universities was also difficult. A modern university faculty relies on income, partly based on projects secured and PhDs completed. Although this is primarily about financial management, the discussions between ARCNL and the universities also reflected different perceptions of science; they did not necessarily share the same definition of good science.

The first issue was to whom the grants were actually awarded. This issue had been brewing since the first ERC Starting Grant was awarded, but it intensified in 2016. In the university's view, several researchers at ARCNL had received grants that they had applied for in their pre-ARCNL years. VU Amsterdam saw an opportunity to claim these sometimes large grants as their in-kind contribution to ARCNL, a proposal that was not warmly received by the ARCNL board.[54] After all, when researchers applied for the grants, the university could not guarantee that this money would become available and therefore could not reserve it as a contribution. As a compromise, those early grants would count as a 50% in-kind contribution from the university.[55]

The discussions about ARCNL's funding also touched on the legitimacy of ARCNL's existence. The universities had begun a cost-benefit analysis of the ARCNL initiative in 2016, and it was increasingly felt that it would turn out to be a loss-making venture.[56] This could have consequences for their commitment to the institute. The researchers might have to raise more money to balance the budget, diverting some of their focus from research to writing grant applications. The VU Amsterdam representative reportedly communicated that ARCNL should be a zero-sum game for the university.[57] Each of the universities contributed €625,000 annually, and they assumed that the science faculties would 'fully recoup their investment through university allocation models'.[58] In principle, it should not cost them any money.

Thus, the universities' (financial) support for ARCNL seemed to be waning.

A modern university faculty relies on income. This income no longer comes directly from the government but is also divided over what are known as second and third revenue streams: indirect government funding via competitive grants and contract-research funding from the industry. In addition, there was the bonus for PhDs obtained at Dutch universities. For each PhD awarded, the faculties received between €65,000 (VU Amsterdam) and €95,000 (University of Amsterdam). Assuming ten PhDs per year, equally divided between the universities, they calculated that they would earn back €7.81 million of their combined contributions (assuming a 90% graduation rate) over a ten-year period. Additionally, a percentage of the grant money received – estimated at €3.6 million – could be counted as overhead for the universities. These financial expectations were crucial to the universities' commitment to ARCNL.

Halfway through 2016, the universities concluded that they would not receive anywhere near this amount of income. They had been counting on 100 PhDs in ten years, a number that seemed higher than ASML or AMOLF had initially suggested.[59] Not only did they fall short of this target, but the PhD bonus was also set to decrease significantly due to a backlash against what critics now saw as a 'perverse incentive' in the scientific community.[60] The universities therefore came to a firm conclusion: a 'financially sound business case for the universities does not seem realistic'. The question now was who was to blame.

In the eyes of university administrators, too little external funding was being secured. This led to too few PhD candidates and too low an overhead contribution that could be charged to the grants secured. Although they had a positive view of the partnership as a whole, the universities saw the new institute as 'a financial problem for the science faculties at the University of Amsterdam and VU Amsterdam', which was caused by 'ARCNL's underperformance'.[61] In the area of education, several things had not been accomplished, although in mutual agreement.

They believed that the master's track in nanolithography had rightly not been established because it would be 'too narrow a specialisation' within the physics programme.[62]

The figures from the universities paint a clear picture of the lofty expectations for ARCNL. For the ten-year period, the total expenditure was expected to be €11.25 million, with total revenue of €12.45 million. By 2016, €3.65 million of this had been obtained, and simulations suggested that the universities would earn a total of €6.68 million in revenue, leaving them with a combined deficit of €5.77 million.[63] 'The worrying conclusion is [...] that the failure to acquire successful projects has destroyed the financial foundation of [...] the business case'.[64]

For the universities, the ARCNL problem was largely a human resources problem. As this chapter has shown, projects were indeed secured by ARCNL tenure trackers, who represented the young institute's main 'revenue generation capacity'. However, they were mostly formally employed by FOM, so their contributions could not be easily accounted for via the universities. The ARCNL board found itself confronted with an 'impossible situation'.[65] The most logical solution seemed to be to allow these grants to be counted in the universities' allocation models in a roundabout way. At the same time, the universities worried that they would appear to no longer be enthusiastic about ARCNL, while all they were trying to do was address what they saw as a genuine financial problem.

The universities could count on little sympathy from the ARCNL board. Their expectations were labelled unrealistic and, what's more, it appeared that not everyone was aware of the universities' financial expectations. According to chair of the board Van Vuren, the universities clearly had an 'over-optimistic view' of the situation, which they presented to the ARCNL directors as a fait accompli.[66] Another problem was that the universities' expectations were in stark contrast to those of ASML. They still hoped that the ARCNL researchers would be able to focus primarily on research, not on obtaining project funding.[67] ARCNL therefore did not see it as its mission to obtain grants ('quite the contrary'). The management expected to have produced around 70 PhD candidates by 2024, of whom around 25 would have to be financed by project funding.[68] This was significantly lower than the total the universities expected.

These expectations were based partly on financial realities, but also on their notion of good science. The universities did not see why ARCNL's 'level of ambition' regarding revenue generation should be lower than that of comparable research institutions. For example, in 2016, the Institute of Physics at the University of Amsterdam obtained approximately 45% of its funding from external grants, and AMOLF around 50%.[69] At ARCNL, this figure was around 15% in 2016.[70] On this point, the universities emphasised the pressure that ASML was putting on ARCNL staff. Was it not good to create more scientific independence from ASML through externally funded projects? According to the universities, ASML was exerting a form of 'pressure not conducive to a healthy academic climate for basic research'. They believed that applying for grants 'in competition with peers' was a sign of a healthy academic climate and should therefore be encouraged.[71] So the ARCNL partners also turned to each other to reinforce their own arguments.

The discussions between the Amsterdam universities and ARCNL show how the funding structures ultimately also affected the participants' perceptions of science. The universities believed that the ability to obtain grants was a defining characteristic of good researchers, while ASML took the opposite view.

Nostalgia for Bell Labs and Philips' NatLab

How could the different visions for ARCNL unite to create a collaborative 'Bell Labs on the Amstel'? More frequent contact between ARCNL and ASML seemed to yield positive results. In early 2017, Martin van den Brink and Jos Benschop – the leading technical experts at ASML – were optimistic. Appreciation for ARCNL had grown rapidly among ASML's senior management. In terms of revenue generation, ASML also finally reached an agreement with the universities. Although ASML continued to believe that conducting research was more important than securing project funding, it recognised that the target of €2.5 million was in line with the original agreements, and it was also desirable to remain on good terms with the universities and to satisfy them in this regard. However, the ARCNL board ultimately decided that, given the institute's start-up phase, it was not realistic to impose this requirement on the scientists immediately. The €2.5 million revenue capacity remained a distant goal for a later phase.[72] The 2017 target was set at €1.1 million; they would only aim for €2.5 million per year from 2020 onwards. However, that did not settle the discussion.

Perhaps the most important voice in the discussion about the revenue generation targets was that of the researchers themselves, and they were not enthusiastic. There seemed to be discontent from the researchers about the stated target of €2.5 million: 'Getting funding is not, and should not be, a goal in itself.' Securing grants was not a problem on its own and it was a common aspect of scientific research, but the motivation for the new target was unconvincing. The researchers felt that ARCNL's revenue generation policy was aligned with the administrative problems of the universities rather than the ambitions of ARCNL itself. In their view, scientists' productivity should primarily be measured by their scientific work and not in terms of 'fundraising'.[73] The board recognised that this view reflected a broader dissatisfaction at ARCNL.[74]

The financial expectations also conflicted with the perceptions researchers had when they joined the ARCNL initiative. The target was perceived as more ambitious than at the universities; the researchers believed that their university colleagues had no concrete revenue targets at all. Even more significant was the contrast with their original motivation for participating in the ARCNL project. In their minds, 'a major reason' for coming to ARCNL was that they would experience less pressure to bring in project funding. This was seen as a 'trade-off' for 'giving up some of their academic freedom' and tailoring their scientific interests to the ARCNL programme. In their eyes, the revenue-generating target cancelled out this original advantage. The researchers felt that they might as well return to the university, where they could work with 'complete academic freedom'.[75] Ultimately, the researchers thought that ARCNL should be an institute where people indeed had to give up some freedom and autonomy in terms of content, but this should be balanced by less guidance and pressure to obtain research grants.

The ideal of a scientific sanctuary – the very image that Bell Labs and NatLab represented – was still a long way off in terms of research. Alongside the discussions about earning back costs, ARCNL had sought closer ties with ASML in 2016. But some did not see this as a positive development. Discussions around the scientific evaluation in 2017 illustrated the differing positions that were taken in this regard. The Scientific Advisory Council's report (which we will discuss in more detail below) was somewhat critical, but Professor Willem Vos (University of Twente) felt it was not critical enough and he did not want his name to be attached to the final report. Allegedly, all the committee members were critical of ARCNL's development, but Vos wanted this to be expressed 'more strongly and explicitly'.[76]

Specifically, Willem Vos wanted to address ASML's influence on ARCNL more forcefully than his colleagues. This was yet another clash between different visions of scientific research. The ARCNL board was aware of Vos's opinion that ASML's influence was 'too strong' and was not conducive to 'good science'. Reportedly, Vos believed that the scientists at ARCNL were 'too sensitive and nervous' about ASML's opinion and were therefore too conservative in their approach to research. Vos believed that the relationship with ASML lay at the root of both the limited scientific output and the narrow focus on generating revenue.[77] By 2016, ARCNL had published only three research papers. ASML's influence was undeniable, but in a sense it also formed the foundation of ARCNL. Furthermore, the need to set up laboratories and experiments led to a logical delay in obtaining scientific results. That is why

Frenken did not agree with Vos that this ultimately undermined the scientific quality. He believed that the three publications reflected a 'silence before the storm'.[78] In the end, Vos would have been alone in the severity of his criticism, as was also evident from his absence in the final report.

The committee recognised that the partnership between ARCNL and ASML had improved considerably and was running more smoothly for a growing number of scientists. Nevertheless, there were still concerns about the way in which the partnership between ARCNL and ASML was taking shape in the scientific programme. Particularly important in this was the role of young scientists who could grow into tenure trackers at ARCNL. The advisers felt that these scientists were 'ARCNL's biggest capital' and the main key to the new institute's success. That a young scientist would now leave due to the sometimes unpleasant interaction showed the critics that the partnership with ASML did not always contribute to 'a stimulating and creative academic research environment'. But Frenken believed that the situation had already improved considerably: ASML was sometimes harsh, but he felt that the number of unpleasant interactions had already been greatly reduced.[79]

To revive the idealised image of Bell Labs, the academic critics felt that greater emphasis should be placed on scientific freedom. The advisers were not blind to the hybrid and therefore complex character of ARCNL, but they also believed in the value of academic freedom in such a structure. To make ARCNL a 'hotspot of inventiveness and creativity', 'harkening back to the days of Bell Labs and the Philips Research Lab', it was necessary to give the researchers 'breathing room', which was considered essential for 'world-class science'.[80] After all, ARCNL was an 'unprecedented' experiment in the Dutch scientific landscape, positioned somewhere between academic blue-sky research and an industrial lab. ARCNL should be given time 'to develop its own unique style, combining the best of the academic and industrial worlds, and making a significant impact in both'.[81] Given patience and a few tweaks, ARCNL would enjoy a successful future.

ASML also recognised that ARCNL was a complex institute because of the wide range of expectations and the need to reconcile 'potentially conflicting objectives'. As this chapter shows, the universities, advisers and researchers all had perspectives that differed from that of ASML. ASML felt that combining 'scientific excellence' with 'serving ASML' may have been a 'stretching' objective, but it was nevertheless a desirable goal.[82] One major criticism was that the ratio of PhDs to postdocs may have been suboptimal. If ARCNL employed more experienced postdoctoral researchers instead of PhD candidates, it might be better able to meet expectations in terms of both publications and technical input. However, this was completely contrary to the basic starting point of the universities, which was partly based on the bonus paid for PhD candidates.[83]

Language and communication took centre stage in the discussions about the partnership with ASML (and they are an aspect of ASML's culture for which the company has become infamous in recent years). To address the scientists' concerns, ASML emphasised in its 2017 evaluation report that it had been formulated 'with care' but was nevertheless intended to provide clarity. According to the report's authors, its directness was not intended to be 'offensive', but 'reflects a normal way of communication within ASML programs'.[84] ASML had delivered a critical report, but according to Schuurmans, the 'interaction had really improved' and the partnership was 'really fantastic' in a few areas, even though there were a few points of concern. Frenken was surprised that the critical report was 'actually meant to be very positive' and would have liked to see the report 'project a more positive attitude'.[85]

Joost Frenken was not pleased with the constant criticism from ASML, and the way it was formulated struck a nerve with some researchers. In 2017, ASML made it clear that they were less interested in ARCNL's processes programme, which focused mainly on the chemical reaction of photoresist to EUV light. Photoresist is a very thin layer of film on which EUV light writes a pattern. However, Frenken countered that ARCNL had already set up groups, recruited researchers, and planned and carried out experiments. If ASML suddenly had less interest in this, they would simply have to accept that they would have less interaction with that group. ASML's comments about their communication style and corporate culture were also met with little sympathy: 'An attitude that reflects "a normal way of communication within [an] ASML program" does not work for ARCNL, simply because *ARCNL is not an ASML program*'.[86] Ottosson's departure was still fresh in ARCNL's memory and the ARCNL board wished ASML had learnt more from that episode.

ASML interpreted ARCNL's attitude as defensive and emotional. This mainly concerned the research group working on photoresists, where the interaction between ASML and ARCNL seemed more strained than with other groups. This was also a sensitive discussion because a tenure tracker had been specifically hired for this subject a year earlier. According to Frenken, the staff became 'frustrated' by ASML's attitude, which they felt undermined the enthusiasm and had a 'demotivating' effect. In the opinion of the ARCNL board, ASML paid too little attention to the personal bond between academic researchers and their projects:

> 'Frenken emphasises that academic researchers feel an extremely strong personal connection to their research, which means that a comment about the usefulness of their research can easily be taken personally. You can't expect a tenure tracker who has just been appointed and is well on his or her way after a year to completely overhaul their scientific plan.'[87]

The discussions eventually made it clear that 'there are serious doubts at ASML about the productivity of the partnership between ASML and ARCNL'.[88] There was little choice but to have senior management address the problems of the young institute. Stan Gielen on behalf of NWO, Jos Benschop on behalf of ASML, and the rectors of the Amsterdam universities had to meet to sort things out. Their verdict was that it was time for leadership: as director, Joost Frenken had to clarify how ARCNL could still be made a success.[89]

The idealised image of a Bell Labs or NatLab was certainly strong, but it could not be translated directly into the Dutch 21st-century context. The degree to which research at ARCNL could be freely designed was an issue that affected both the academic and industrial research culture, but a shared perception of science for ARCNL did not yet exist and had to be developed as they went along.

From FOM to NWO

A final crucial difference between ARCNL and a Bell Labs or NatLab was the involvement of the national science funder, the Dutch Research Council (NWO). It was precisely because of its involvement and ARCNL's position as an NWO institute that ARCNL was not allowed to be purely an ASML institute. In parallel with the discussions between ARCNL, ASML and the universities, it was therefore important for ARCNL to play a national role in the Dutch scientific landscape as an NWO research institute, and it was precisely in this area that significant changes were imminent.

ARCNL had been born out of FOM, but in 2014 it became clear that the physicists would have to give up their distinct status in the Dutch scientific establishment. FOM would be fully absorbed into NWO and would no longer exist as a separate entity. This also had consequences for ARCNL. When it was founded, it had in a sense still been able to benefit from FOM's autonomy. For example, the first instalment of Industrial Partnership Programme (IPP) funds was obtained through what was called an 'IPP-light' procedure: only the FOM board was supposed to decide on this, and it concerned funding for 2014. Only after that would a full proposal, including referees and peer review, be submitted for a five-year term.[90] The FOM board ultimately released the funds for 2014 and 2015 as a single payment (€3 million in total), and it was assumed that ARCNL would develop a full IPP proposal later.[91] The short lines of communication at FOM thus gave ARCNL the financial flexibility it needed in its early years.

The relationship between FOM and NWO had always been complex. FOM was often characterised as a flat organisation where researchers were in charge of their own money and had little use for bureaucratic procedures. In contrast, NWO tended to be seen as old-fashioned and slow, 'an untidy patchwork'.[92] Under the Rutte II cabinet (2012–2017), the idea was developed that the NWO organisation should be less rigid and compartmentalised, and it should focus on societal challenges instead of disciplinary divisions. All the various governing boards and area boards – including that of FOM – had to be abolished. A member of FOM's board, Unilever's head of R&D Rob Hamer, believed that the loss of FOM would represent the loss of 'a pearl' of Dutch science.[93]

Many people saw FOM as a special organisation. In 1996, an international panel of physicists described it as a 'physicists' organisation, run by the researchers for the researchers', in which the leading role was played by the biggest names in Dutch physics.[94] The fact that FOM both distributed research funds and managed institutes was seen as one of the strengths of Dutch physics. The experts concluded that 'many other countries envy the Netherlands for having an organization such as FOM. It is hoped that FOM will be able to continue its work in the same spirit for many years to come. FOM money has always been well spent money.'[95] So FOM received very positive evaluations from physicists.

The 2014 reform did not come out of nowhere. From the end of the 1990s, NWO made increasingly serious attempts to fully integrate FOM into its organisation. Historian of science Dirk van Delft described it as 'a frontal attack on FOM's independence'.[96] Under the leadership of Hans Chang, FOM had always managed to fend off these attacks. Chang was less a famous physicist than a physicist who had ended up in that position due to his involvement in policy, and he acted accordingly.[97] His successful defence of FOM earned him the nickname 'the Indiana Jones of Dutch science': he had managed to keep the rope bridge connecting NWO and FOM in place and to repair that bridge over a deep ravine.[98]

However, starting in 2014 it became clear that NWO and all the area boards would undergo thorough reform. The traditional division into disciplines and area boards was rendered obsolete by the new top sectors policy, which focused primarily on overarching themes.[99] There was a need for an organisation that had a vision of the larger themes, and one in which the large research institutes, in particular, would be better connected to the research domains.[100] ARCNL would not escape this dynamic either.

It quickly became clear to those involved that NWO's transition would have consequences for ARCNL. The institute's continued existence depended not only on the enthusiasm of FOM, ASML or the universities, but also on the extent to which it was integrated into the national science system. The NWO reform meant that the institutes also had to be equipped with an Institute Advisory Board that could expand the institute's network, particularly at the political level. Looking ahead to 2018, a broader evaluation of the Dutch institute landscape was also on the agenda, and such a new advisory board was probably 'strategically desirable' to ensure ARCNL's position.[101] This board could both provide a 'PR network' and assist in 'lobbying'.[102] During the same phase, ARCNL attempted to gain definitive recognition as a regular institute within NWO. A full transition to NWO would also mean that ARCNL would have to raise its profile more broadly in the Dutch scientific landscape and could no longer afford to be seen as an Amsterdam-based ASML institute.

Conclusion

In retrospect, the years from 2013 to 2017 were the 'infancy' or 'baby phase' of ARCNL.[103] It was a phase of enthusiastic growth, but also one that was filled with growing pains – both physical and mental. Accommodation was organised, laboratories were set up, an organisation was put together and funding was secured to balance the budget. At the same time, the partners involved also had to work together on the direction and organisation of the research. How did those involved build what some described as 'Bell Labs on the Amstel'?

The establishment of a completely new physics research institute went through fits and starts. The partners all had different perceptions of science, organisational cultures and expectations for ARCNL. The period between 2014 and 2017 was characterised by the recruitment of group leaders and the establishment of research groups, but also by the growing mutual irritation between the institutions involved. When would ARCNL be considered a success? And what, in fact, would be considered a sign of success? The answers to these questions were not obvious and form the theme of the final chapter.

Notes

1. Governing Board, report of meeting 43, 6 April 2017. ARCNL archive.
2. Parikka et al. *The Lab Book: Situated practices in media studies* (University of Minnesota Press, 2022), 187–212.
3. Jon Gertner, *The idea factory: Bell Labs and the great age of American innovation* (Penguin, 2012), 1–2; Mark Raizen, 'Commentary: Let's re-create Bell Labs!', *Physics Today* vol. 71, no. 10 (2018), 10–11; Iulia Georgescu, 'Bringing back the golden days of Bell Labs', *Nature Reviews Physics* 4, 2 (2022), 76–78.
4. Parikka et al. *The Lab Book*, 187–212.
5. ARCNL Governing Board, report of meeting, 5 December 2013. ARCNL archive.
6. 'Draft report of chair's meeting', 12 February 2014, NA, NWO Archive, 2.25.109, 1669.
7. 'Grafeen, een wonderbaarlijk materiaal, lost grote belofte nog niet in' ['Graphene, a miracle material, does not yet fulfil its great promise'], *Het Financieele Dagblad*, 14 June 2014; 'Wondermateriaal van vuilniszakkenkwaliteit' ['Wonder material of bin-bag quality'], *de Volkskrant*, 29 June 2013.
8. 'Acute vragen ASML ook bij instituut' ['Burning questions for ASML, also at institute'], *Eindhovens Dagblad*, 8 November 2013.

9. Message from Jos Benschop to Wim van Saarloos, 22 January 2014, ARCNL archive.
10. Message from Hendrik van Vuren to ARCNL stakeholders, 'ARCNL: Partnership agreement and special meeting of the Governing Board on 3 February 2014 at 09:00', 24 January 2014. ARCNL archive.
11. Frenken to the ARCNL Governing Board, 'RE: my own research on graphene growth', 28 January 2014, ARCNL archive.
12. Ibid.
13. Joost Frenken to Hendrik van Vuren, 29 January 2014. ARCNL archive.
14. Message from Joost Frenken to Vinod Subramaniam and Wim van Saarloos, 21 March 2014. ARCNL archive.
15. ARCNL Governing Board, report of meeting on 28 February 2014. ARCNL archive.
16. ARCNL Governing Board, report of meeting no. 19, 4 September 2014. ARCNL archive.
17. Message from Wim van Saarloos, 28 February 2014, ARCNL archive.
18. Ibid.
19. Message from Hendrik van Vuren, 27 February 2014. ARCNL archive.
20. Ibid.
21. ARCNL Governing Board, report of meeting no. 15, 6 March 2014. ARCNL archive.
22. Ibid.
23. Memo from Huib Bakker, 'Summary of visit to ASML + structure and programme for INL', 2 March 2013. AMOLF archive.
24. Ibid.
25. ARCNL Governing Board, report of meeting no. 22, 8 January 2015. ARCNL archive.
26. Message from Frenken to Hendrik van Vuren, 7 January 2015. ARCNL archive.
27. Frenken and Reijnders, 'ARCNL philosophy', 12 March 2015; ARCNL Governing Board, report of meeting no. 24, 5 March 2015. ARCNL archive.
28. ARCNL Governing Board, report of meeting no. 24, 5 March 2015. ARCNL archive.
29. ARCNL Governing Board, report of meeting no. 26, 7 May 2015. ARCNL archive.

30. Hendrik van Vuren to Karen Maex, 'Working arrangements between ARCNL-UvA-VU on the management of personnel and operations in relation to the in-kind contributions of the universities', 24 June 2015. ARCNL archive.
31. Joost Frenken, 'Response to 2015 SAC report', 27 January 2015. ARCNL archive.
32. *ARCNL SAC Report*, 2015. ARCNL archive.
33. Message from Hans de Groene, 'Strategic partnerships meeting 10 Feb', 29 January 2015. ARCNL archive.
34. *ARCNL SAC Report*, 2016. ARCNL archive.
35. Ibid.
36. Joost Frenken, 'Response to 2016 SAC report', 7 February 2016. ARCNL archive.
37. ARCNL Governing Board, report of meeting no. 29, 1 October 2015. ARCNL archive.
38. Ibid.
39. Report on ASML-FOM Steering Committee (i34-FEUL), 3 November 2015, NHA, FOM Archives, 449, inv. no. 1566.
40. Message from Niek Lopes Cardozo to Hendrik van Vuren and Christa Hooijer, 'Conversation with Jos Benschop', 2 November 2015. ARCNL archive.
41. Message from Christa Hooijer to Van Vuren, Van den Hout, De Witte, 'Brief report on conversation with Joost Thursday evening', 7 November 2015. ARCNL archive.
42. ARCNL Governing Board, report of meeting no. 30, 5 November 2015. ARCNL archive.
43. Wim van der Zande (with input from others), 'SWOT analysis report connected to the ARCNL-ASML Conference of April 20', ARCNL archive.
44. ARCNL Governing Board, report of meeting no. 29, 1 October 2015. ARCNL archive.
45. ARCNL Governing Board, report of meeting no. 30, 5 November 2015. ARCNL archive.
46. Frank Schuurmans & Wim van der Zande, 'Executive Reaction from ASML based on the ARCNL-ASML Conference of April 20', ARCNL archive.
47. Joost Frenken to ARCNL's Governing Board, 'ASML's executive reaction, ARCNL-ASML conf.', 20 June 2016. ARCNL archive.
48. 'SWOT analysis report connected to the ARCNL-ASML Conference of April 20', ARCNL archive.

49. Joost Frenken to ARCNL's Governing Board, 'ASML's executive reaction, ASML-ARCNL conf.', 20 June 2016. ARCNL archive.
50. Ibid.
51. ARCNL Governing Board, report of meeting no. 40, 3 November 2016. ARCNL archive.
52. Ibid.
53. Ibid.
54. ARCNL Governing Board, report of meeting no. 33, 4 February 2016. ARCNL archive.
55. Ibid.
56. ARCNL Governing Board, report of meeting no. 35, 7 April 2016. ARCNL archive.
57. ARCNL Governing Board, report of meeting no. 36, 12 May 2016. ARCNL archive.
58. ARCNL Governing Board, report of meeting no. 37, 7 July 2016. ARCNL archive.
59. Ibid. In January 2013, Bart Noordam spoke of an annual 'outflow' of 6 PhDs a year. Cf. Chapter 2.
60. Martijn van Calmthout, 'Kabinet snijdt in hoge promotiebonussen op universiteiten' ['Cabinet cuts high PhD bonuses at universities'], *de Volkskrant*, 25 November 2014.
61. ARCNL Governing Board, report of meeting no. 37, 7 July 2016. ARCNL archive.
62. ARCNL Governing Board, report of meeting no. 38, 1 September 2016. ARCNL archive.
63. 'Evaluation of ARCNL business case – explanation and rationale for financial evaluation', appendix to ARCNL GB, 6 October 2016. ARCNL archive.
64. Ibid.
65. ARCNL Governing Board, report of meeting no. 38, 1 September 2016. ARCNL archive.
66. ARCNL Governing Board, report of meeting no. 39, 6 October 2016. ARCNL archive.
67. Ibid.
68. 'Update to memo on ARCNL revenue generation capacity', 2 December 2016. ARCNL archive.
69. *Evaluation report, Amsterdam University Physics* (February 2018), 26; *AMOLF Evaluation 2011–2016 – Physics of Functional Complex Matter* (February 2018), 11.
70. *ARCNL Evaluation 2014–2016 – Advanced Research Center for Nanolithography* (November 2017), 14.

71. 'Position of UvA and VU on memo on ARCNL's revenue generation capacity', appendix to ARCNL GB meeting of 12 January 2017. ARCNL archive.
72. ARCNL Governing Board, report of meeting no. 41, 12 January 2017. ARCNL archive.
73. Stefan Witte, 'Re: memo earning capacity ARCNL Staff' 27 February 2017. ARCNL archive.
74. ARCNL Governing Board, report of meeting no. 42, 9 March 2017. ARCNL archive.
75. Ibid.
76. Joost Frenken, 2017 SAC Report, response', 3 March 2017. ARCNL archive.
77. ARCNL Governing Board, report of meeting no. 42, 9 March 2017. ARCNL archive.
78. Joost Frenken, '2017 SAC Report, response', 3 March 2017. ARCNL archive.
79. Ibid.
80. 'ARCNL SAC Report', 2017. ARCNL archive.
81. Ibid.
82. Wim van der Zande and Frank Schuurmans, 'ASML's ARCNL Evaluation Report', 17 March 2017. ARCNL archive.
83. Joost Frenken, 'Response to ASML Appreciation Report 2017', 31 March 2017. ARCNL archive.
84. Wim van der Zande and Frank Schuurmans, 'ASML's ARCNL Evaluation Report', 17 March 2017. ARCNL archive.
85. ARCNL Governing Board, report of meeting no. 43, 6 April 2017. ARCNL archive.
86. Joost Frenken, 'Response to ASML Appreciation Report 2017', 31 March 2017. ARCNL archive.
87. ARCNL Governing Board, report of meeting no. 43, 6 April 2017. ARCNL archive.
88. Stan Gielen, 'Summary of conversation between Gielen, Subramaniam, Maex and Benschop', 25 April 2017. ARCNL archive.
89. Ibid.
90. Message from Joost Frenken, 'Extra section for GB meeting on Thursday 5 June, Start IPP proposal', 2 June 2014. ARCNL archive.
91. 'Documents for your meeting on 1 July 2014; IPP application "physics for nanolithography" (no. i42, PNL)', 24 June 2014. ARCNL archive.

92. Marcel aan de Brugh, 'Van lappendeken naar waterhoofd' ['From patchwork to top-heavy'], *NRC Handelsblad*, 27 November 2014.
93. Ibid.
94. VSNU, *Quality assessment of research – An analysis of physics in the Dutch universities in the nineties* (Utrecht, 1996), 14.
95. Ibid., 15.
96. Van Delft et al., *Snaren, spiegels en plakband – 70 jaar Nederlandse natuurkunde* [*Strings, mirrors and sticky tape – 70 years of Dutch physics*] (W-Books, 2017), 166.
97. Ibid., 166–167; Also see: Hans Chang & Dennis Dieks, 'The Dutch output of publications in physics', *Research Policy* vol. 5, no. 4 (1976), 380–396.
98. Van Delft et al., *Snaren, spiegels en plakband* [*Strings, mirrors and sticky tape*], 173; FOM *Express,* vol. 22, no. 2, October 2009, 4.
99. Van Delft et al., *Snaren, spiegels en plakband* [*Strings, mirrors and sticky tape*], 201–202.
100. Ibid., 206–207.
101. ARCNL Governing Board, report of meeting no. 50, 12 March 2018. ARCNL archive.
102. Joost Frenken Memo to GB, 'First proposal for an Institute Advisory Board (IAB) for ARCNL', 17 April 2018. ARCNL archive.
103. Joost Frenken, 'Questions to Stakeholders and ARCNL Staff', 14 June 2017; Joost Frenken to Stan Gielen, Jos Benschop, Karin Maex, and Vinod Subramaniam, 'Towards a New ARCNL', 12 March 2018, ARCNL archive.

4) Towards a new ARCNL

'Blunt behaviour of chipmakers leads to tensions'.[1] In September 2018, the NRC *Handelsblad* newspaper published an article with this headline about the difficult collaboration between ASML and ARCNL. The article was part of a series about the alleged influence of external funders on scientific research, and it took a critical look at the role of the industrial community. Wherever science and entrepreneurship meet, questions about autonomy, freedom and independence always arise. The newspaper's approach is typical of the binary perspectives on the phenomenon of the entrepreneurial scientist. Are they heroes who help the economy by applying science to society's needs or are they opportunists who corrupt pure science and put profit before social interests?[2] Commentators associate the differences between academic and industrial research with sharp institutional, intellectual and even moral divisions. Yet it is precisely this perspective that clouds our view of the actual situations of the scientists who work at this crossroads.

We know very little about the concrete reality of public-private partnerships in physics. How do scientists and administrators perceive themselves and their environment? What are the dilemmas, contradictions and challenges faced by the entrepreneurial scientist? The period between 2017 and 2021 was a crucial phase in the formation of ARCNL. While cynicism still prevailed in 2017, by 2021 director Joost Frenken was a staunch advocate of the public-private partnership model as had been achieved at ARCNL. Basic and autonomous science could go hand in hand with industrially relevant research, he believed.[3] There clearly had been a turnaround in the way ARCNL worked between 2017 and 2021.

The phase leading up to this steady state in which ARCNL combined its intended size with an increasingly clear way of working provides some understanding of the public-private partnership between the universities and ASML. Contrary to what the headlines suggested, the partnership was complicated by more than just perceived tension between basic research and ASML's industrial interests. As the previous chapter showed, the public partners also had differing perspectives. How is it that the opinions of those involved changed so dramatically in such a relatively brief time? And what can we now regard as the steady-state model of ARCNL? These questions are the focus of this final chapter as we search for the 21st-century entrepreneurial scientist.

Matrix VII at Science Park

The links between entrepreneurship and science became visible in the spatial design of universities in the Netherlands from the 1980s onwards. Inspired by the world-famous Silicon Valley, science parks sprang up like mushrooms around Dutch university towns.[4] The Silicon Valley dream was based on the idea that the presence of (small) technology companies, along with universities of applied sciences and research universities in a single location, would lead to spontaneous and creative exchange, and thus to innovation and economic development. Universities were no longer just for students. 'In addition to students, businesses should also be drawn to the campuses', journalists noted in 2015.[5]

In the form of Matrix Innovation Centres, business incubators were established at Amsterdam Science Park, and ARCNL found a permanent home there. That initiative was a collaboration between the University of Amsterdam, the city of Amsterdam, NWO and Rabobank that launched in 1989. Since 2022, VU Amsterdam has also been part of the alliance.[6] The Matrix Innovation Centres were intended to provide facilities for both companies and academic start-ups.

5. An inside view of ARCNL. Photo: Wouter Jansen.

On 6 June 2017, the first pile was driven into the ground for the seventh Matrix building, which would eventually be home to ARCNL and several other start-ups. Starting in December 2018, ARCNL moved from the temporary office buildings next to AMOLF to the new location.[7] Finally having its own building was a major step in the further independence of the young institute. In 2015, it had already moved from AMOLF premises to temporary buildings, symbolically cutting the umbilical cord between the two institutes. Prior to that, ARCNL had also used NIKHEF's PiMu building (which had been left over from the days of nuclear physics experiments) for experimental spaces.

The new building was constructed in a modern style and looked something like an 'industrially designed hotel lobby'.[8] The building caught the attention of a local architecture journalist from newspaper *Het Parool*. In his opinion, it was primarily functional and safe but not very spectacular in its design. Matrix VII is covered in glass – a number of rooms are visible to outsiders, but many others are not. The journalist thus described Matrix VII as a paradoxical building, 'whose facade is almost entirely made of glass and yet cannot tolerate daylight'.[9] Daylight must not interfere with the many experiments conducted at ARCNL, which is why the building is, in a sense, constructed as 'a box with a ring around it'.[10] This design created space for an open atrium in the centre of the building, surrounded by offices. The journalist thought that the whole building fit in well with the 'university enclave', and he noted that Matrix VII reinforced the 'nerdy' feel of the campus.[11]

The journalist's critical view was in keeping with the history of Amsterdam Science Park. By no means was everyone persuaded by the philosophy behind science parks, and this criticism also applied here. According to the then director, Amsterdam Science Park was a loss-making venture in the 1990s.[12] A study of the Dutch 'campus economy' concluded that in 2014, Amsterdam Science Park was still less successful than its counterparts elsewhere.[13] A critical article in *De Groene Amsterdammer* magazine was not very enthusiastic about the new initiatives, including ARCNL:

> 'Universities are transforming their campuses into glorified business parks and tailoring their research work to the wishes of the government and companies in the top sectors. The hope is that this will open the floodgates and allow private euros to flow into the academic world. The reality is more intractable: it is the government that pays and the university that bears the risks.'[14]

In the eyes of these journalists, the founding of ARCNL was a prime example of this dynamic at Dutch universities: ASML's investment 'seems like a lot, but the majority of the investment is public money'.[15] The journalists saw ARCNL as an example of the moderate success of Dutch innovation policy. They believed that while the government invested a lot of money in public-private partnerships, companies still contributed less than had been hoped.

A failed compromise: ARCNL 1.1

The critical article published in *NRC Handelsblad* in summer 2018 seemed to come as a shock, but in fact it was nothing more than a belated scoop. The first squabbles at ARCNL were already apparent in 2015, and the biggest problems had just been resolved by the time the article was published. ARCNL director Frenken had given the institute's researchers carte blanche to talk to the journalist about the goings-on at ARCNL. Although some researchers shared their critical opinions with the newspaper, others countered that there was 'enough space for basic research' and they saw ARCNL's growing pains as a normal part of a new and ambitious initiative.[16] Not all researchers experienced the partnership with ASML in the same way.

Nevertheless, the article was published after what can be seen as a crucial phase in the history of ARCNL. In addition to the annual evaluations by the scientific advisers (SAC) and ASML, ARCNL was subjected to a wide-ranging evaluation in 2017 that NWO organised for all its institutes every six years. The evaluation committee, which consisted of Dutch and international experts, had a strong message for everyone involved: ARCNL was a very complex organisation, and this complexity had three causes: '[It] is introduced by financial pressures from the universities, mission-related stress from AMOLF and impatience from the side of ASML'.[17] The assessors were especially critical of the universities and ASML: ARCNL should not be pushed to bring in more research funding, but neither should it be micromanaged by its industrial partner.[18] In the committee's view, 'this tension [was] a threat to the Centre's viability which needs to be addressed'.[19] The Amsterdam universities and ASML in particular should rethink the way they dealt with ARCNL.

Halfway through 2017, it had become clear to all involved that ARCNL could not be expected to simultaneously perform at the highest scientific level, produce patents and inventions, and secure grants and facilitate education. In other words, ARCNL could not be AMOLF, TNO *and* a university all rolled into one. Furthermore, ARCNL could not be seen as a regular NWO institute, nor as an extension of the ASML research department.[20] To ensure the future of ARCNL, the partnership needed to be thoroughly overhauled based on clear and shared expectations.[21]

So what could be expected from ARCNL? To define the new set of expectations, all the institute's partners and staff were invited to take part in an extensive survey. The expectations were tested against various criteria that together painted a clear picture of the tensions around ARCNL. Where did ARCNL fit in between AMOLF and TNO, institutes that represented the extremes of basic and applied research in Dutch science? What was the ideal mix of contract vs. free research in relation to ASML? How many inventions, papers and grants could reasonably be expected? What was the ideal ratio between postdocs and PhD candidates, and to what extent should they be involved in university teaching? These questions would lead to the main questions: 'When will ARCNL be a success? How do you feel about continuing it? What will be lost if it closes?'[22] Nothing less than the survival of ARCNL was at stake:

> 'ARCNL is built on significant investments in human capital ("blood and skin"), in research infrastructure ("bricks and steel") and in prestige ("pride and honour"). It is an experiment that all stakeholders and participants are motivated to provide with maximum support and that simply *has* to succeed. In "Apollo 13 terminology": *Failure is not an option!*'[23]

The new vision for ARCNL was to become a blueprint that came to be known as ARCNL 2.0. The management had explicitly opted for a 'carefully balanced compromise' in which the wishes and requirements of all partners were respected. This included the ambitions of ARCNL's researchers. The management resolved an essential issue – the autonomy of the research – by reserving a third of the research budget for projects that related to nanolithography but might not yet be fully appreciated by ASML. This should satisfy the group leaders' desire for 'sufficient "intellectual" diversity to be an attractive workplace'.[24] The researchers were enthusiastic: they believed the blueprint 'captures the soul of the center' and had succeeded in finding harmony between the stakeholders. Anticipating a positive outcome, they assumed that 'the staff has the expectation that it will be supported by all the stakeholders'.[25] This proved to be wishful thinking.

It was not entirely surprising that ASML had very different feelings about the new proposal. Upon reading it, they expressed a 'sense of disappointment'.[26] In their view, the ARCNL board seemed to be primarily aiming for a compromise in which 'all stakeholders [...] are asked to give in', but with no clear internal changes at ARCNL. ASML would have preferred to see a substantial reduction in the complexity of the institute, something they considered necessary to allow the scientists to focus on their research. They also hoped that the researchers would show more genuine interest in the challenges faced by ASML: the director had to create a 'mindset' at ARCNL in which 'all employees of ARCNL at all times can explain the relation and connection of their research with ASML'.[27] Moreover, others had the impression that ASML was so fed up with the attitude of the universities that they wanted to explore the option of continuing without them altogether, or possibly finding different universities.[28] In any case, ASML was not persuaded by the ARCNL management.

The search for a compromise was not well received by the universities either. Thinking in terms of compromise also implied an approach in which contradictions took centre stage. Instead, the universities were hoping for more emphasis on 'the search for added value in the partnership by joining forces'.[29] Although Frenken's plan accurately reflected ARCNL's existing weaknesses, it lacked a real long-term vision. In the opinion of the deans, it was more of a 'version 1.1. than a version 2.0' of ARCNL, an opinion that was also shared by FOM. Unlike ASML, the deans emphasised the importance of a solid research programme that would generate enthusiasm among the scientists. They further emphasised the importance of 'ambiguity tolerance' for all stakeholders: it was inevitable that an uneasy balance would have to be found between ARCNL's freedom and the involvement of all parties. It was up to Frenken to define how that could be achieved.

NWO, the final partner on the board, was also highly critical of the proposed plan. They saw it as a collection of 'patchwork and compromises' that were incapable of meeting all ARCNL's challenges. NWO believed ARCNL played a 'pioneering role' as an 'institute of the future'. It was a 'crown jewel of the government's top sectors policy' and partly for that reason, NWO was keen to bring it to a successful conclusion.[30] That would require a 'fundamental' change. First, the mission needed to be defined more specifically. What was the goal of ARCNL and how could it be achieved? That was a fundamentally different question from how to satisfy the stakeholders. Moreover, NWO chair Stan Gielen argued for lump-sum financing: that alone would give the ARCNL director the much-needed freedom. Previously, Gielen had stated that he had the impression that Frenken was mainly focused on 'postponing the difficult discussion'.[31] Frenken felt there was no room for a 'completely fresh and new plan' and that it was indeed a 'compromise proposal'.[32]

It was difficult to bring the critical perspectives together, and initially this led to a negative spiral. By the end of 2017, little remained of ASML's original enthusiasm, and the universities were also expected to be more vocal in their support of basic research. 'He who pays the piper calls the tune' was the cynical conclusion among ARCNL's leadership.[33] Was it really impossible to work with 'two cultures' in 'one house'? The original 50-50 distribution of votes on the board between public and private parties was now also seen in a negative light. ASML paid less than half of the total budget but had an equal vote. The fact that ASML had said it was no longer interested in all lines of research was also very critically received at ARCNL and caused a lot of frustration. This was even more frustrating because ASML had agreed to the plans at an earlier stage. The idea that ASML could change its mind in so short a time was extremely uncomfortable.

The desired balance between freedom and involvement was far from being achieved and still leaned strongly in the direction of too much involvement from all the partners. In Frenken's perception, ASML mainly wanted to see 'obedience' combined with a 'businesslike determination' in the implementation of policy and the divestment of research. In addition, the director felt that ASML wanted a lot of insight into and influence on the strategy: 'they want to look over my shoulder and help steer the ship'.[34] The pressure from ASML was also keenly felt. Yet the universities wanted the same obedience in terms of revenue generation. Frenken felt that both sides believed that ARCNL was 'disobedient'.[35] In the eyes of the management, ARCNL could no longer do anything right.

Farewell to a line of research

Much of the discussion between ARCNL and ASML revolved around the different time scales in academic and industrial research. ASML felt that ARCNL moved too slowly when it came to the pace at which changes could be implemented in the scientific programme. The company had stated that 'ASML wants to be able to discuss new research groups with an onset time of 1 year'. Of greater concern was that they also wished to 'discuss the ending of research groups preferably within a period of 2 years'.[36] This mainly concerned the photochemistry research groups and the photoresist activities, both of which were part of ARCNL's broader line of research into processes. Photoresist research focused mainly on new photoresists on which EUV could apply patterns. A new tenure tracker had been recruited for this research in February 2016.

It was this line of research that reflected the difficult struggle in the ARCNL partnership, so it was not surprising that the subject was covered extensively in the national press.[37] At the joint conference between ASML and ARCNL in 2016, it also seemed that this line of research was where discussions on applied and basic research seemed particularly sensitive, while other research groups could already count on more support from ASML. In 2017, it became clear that ASML believed that few relevant research results had been generated, and this was ultimately related to the fact that 'resist is not an ASML primary product'.[38] ASML mainly worked on the machine that produced the light, and the material that was exposed to the light was, strictly speaking, outside their portfolio. ASML admitted that their comment was related to more general concerns: although ASML had initially signalled that they found resist research interesting, they now admitted that 'a growing insight at ASML questions this original perspective'.[39] ASML had changed its position and questioned the role of the new research group at ARCNL.

ASML was accustomed to quickly adjusting research, changing its direction or even phasing it out entirely if deemed necessary. Academic research was structured around time scales that were geared to the cycle of PhD candidates and tenure-track researchers: at least four years on a specific topic. ASML now wanted to see that ARCNL could be agile and dynamic during those four years. ARCNL leadership saw these expectations as an ultimatum: among the management it was felt that ASML was 'holding a knife to [our] throat'.[40] The problem with the research group was not that it was too industrial or too academic in its orientation; on the contrary, the group could count on interest from both angles. However, the industrial interest did not come from ASML, but from other companies. Thus, the dismantling of the group had little to do with the quality of the research and everything to do with ASML's interests. Finally, in early 2018, it was decided that the research topic would be phased out, mainly because the theme 'does not belong to the core activities of ASML'.[41] In the opinion of Joost Frenken, it was a choice between two evils for the ARCNL management: if ARCNL did not phase out the line of research, the very likely implication was that ASML would pull the plug on ARCNL as a whole.[42]

Towards a new ARCNL

Through the end of 2017, ARCNL also regularly seemed to think from a conflict perspective. In this, the entrepreneurial scientist was mainly an uncomfortable compromise, where academic excellence and industrial impact existed alongside each other rather than together: as a group leader was quoted in NRC *Handelsblad*, 'it is difficult to determine what your priorities should be'.[43] But at the start of 2018, the director and the chair of ARCNL's board both began to change their way of thinking. The toxic atmosphere had to end: 'The negative spiral in which all stakeholders complain about each other and about ARCNL must now be broken once and for all [...]. Let's be proud of what has been achieved so far and, above all, let's look forward with that positive spirit.'[44] Joost Frenken was given one more chance to develop a new plan for ARCNL, now also involving the university rectors. On the one hand, this showed how seriously the situation was deteriorating. On the other hand, it was an opportunity: rectors had greater authority and Frenken wanted to explore the more radical options with them. The goal: 'a single management structure like AMOLF has'.[45] Instead of finding a compromise to resolve the conflicts, Frenken now wanted to 'embrace both parties fully'.[46]

Finding a compromise was not enough to ensure a future-proof institute; new guiding principles would also have to be defined. One priority for this was that ARCNL should also be open to other private and public parties. This was not only important for anticipating a possible decrease in contributions from the universities in Amsterdam, but even more so to change the relationship with the private partner. Frenken saw 'the monopoly of ASML as one of the "flaws" of our ARC'.[47] A second priority was to gain control over the sometimes unpredictable impact of ASML on the research programme. It was essential that ASML not be able to 'intervene directly and at random times' in scientific projects. If this were to happen, ARCNL would immediately lose its 'appeal to top researchers'.[48] To address this, Frenken wanted a 'breeding ground' at ARCNL: a place for 'totally unrelated research'. Projects could spend several years in this breeding ground with no immediate assessment of their relevance to ASML.[49]

At the same time, efforts were also made to adopt ASML's way of working. A recurring criticism was that ASML felt ARCNL was adhering too strictly to the original plan as described in the 2013 bid book. According to Frenken, the prevailing opinion in

Veldhoven was that 'ARCNL was sticking too rigidly to the research path it had taken in the past' and ARCNL scientists were reluctant to adapt to the 'dynamic landscape' of application-related issues.[50] Research projects and doctoral programmes worked on 'academic time scales' and could not be changed abruptly. To satisfy ASML's wishes, the ARCNL of the future had to be 'powerful' and 'agile'. They decided to adopt an annual evaluation cycle, which would allow for continuous coordination 'to maintain optimal alignment with ASML'.[51] This dynamic also allowed for new research groups to be started and existing groups to be phased out. That made the measure particularly appealing to ASML. As a result, the rhythm of academic science became more aligned with the rhythm of industrial research.

To better involve the universities in ARCNL, the academic results of their investment had to be made more visible. Prior to this, ARCNL had not been seen as part of the faculties' *'own* academic activities'; 'researchers often operated under the radar', while the financial contribution was a heavy burden on faculty budgets.[52] When ARCNL was set up, in the interests of autonomy, as many staff members as possible were seconded to NWO/FOM. Now the reverse was happening: the intention was that all staff members would be employed by the University of Amsterdam and VU Amsterdam, and all PhD candidates and postdocs would be appointed mainly by NWO. This meant that successes such as obtaining grants would also be successes for the faculties and not just for ARCNL. The aim of this move was to make ARCNL a less complex organisation. By giving the universities more responsibility, ARCNL would be able to focus on its core activities.

Yet even this approach was not immediately embraced by everyone. Academic colleagues who evaluated the institute annually were very suspicious of attempts to align the scientific programme more closely with ASML. They felt there was still a problematic tension between academic freedom in research and ASML's industrial interests. In addition, the relationships as outlined in the new plan were no guarantee of a 'healthy relationship between ARCNL and ASML'.[53] According to the advisers, the proposed annual evaluation programme that was now being set up with ASML was a risk: the coordination procedure should not become an annual review. Changes made too quickly would 'severely compromise ARCNL's ability to achieve its expected scientific excellence'.[54] NWO received particular criticism. Specifically, NWO should have taken on the role of 'watchdog' on the ARCNL board and guarded the independence of the research. The fact that an entire line of research had been discontinued 'well before they have had the chance to demonstrate their full research excellence and impact' was something for which NWO could be blamed.[55]

The decision to dissolve the research group working on photoresist was the most visible and impactful change implemented in the first ten years of ARCNL. The initiative to dissolve the group came from ASML, even though the scientific committee saw the research as extremely relevant for future EUV lithography. Was this not exactly the type of research for which ARCNL was founded? After all, were Bell Labs and NatLab not the inspiration for a research institute where scientists could 'pursue their research with a high degree of academic freedom'? The critics felt that the announced dismantling of a successful research group, including dismissal of the group leader, was completely at odds with their idealised image of ARCNL. Moreover, it was the second example of a tenure track researcher leaving the institute. As such, the critics felt it represented a dynamic 'where the future of a young researcher is adversely affected by the association of ARCNL with ASML'.[56] They would have preferred to see ARCNL work more closely with AMOLF, thus creating more distance from ASML. The SAC had hoped this would lead to more academic science, not less.

Nevertheless, the advisers were alone in their criticism and the partners involved were convinced of the usefulness of working more closely together. ASML had disregarded the SAC's advice, and the ARCNL management was somewhat embarrassed by the committee's 'tough message'. The board had tried to temper the advisers' intense reactions by pointing out the positive perspectives the partners had about the chosen path.⁵⁷ To the ARCNL board, the report felt like 'too little, too late' since all parties involved were already close to a solution.⁵⁸ In a sense, the same was true for the *NRC* journalist: ARCNL had managed to overcome the problems and Frenken had let the staff speak freely with the journalist. Van Vuren, the outgoing chair, had a positive outlook on the future: 'The crop is looking good, so with commitment, dedication and cooperation, a rich harvest lies ahead.'⁵⁹

A future-proof institute

The reform of ARCNL was partly about the question of whether it could and should be more than just an Amsterdam-based ASML institute. One of Frenken's priorities for the new ARCNL was to expand the institute to include more public stakeholders. This would allow it to forge closer links to other universities. This desire was not an isolated one: ARCNL needed to present itself more as a national institute than an Amsterdam one in view of the upcoming evaluation by the Royal Netherlands Academy of Arts and Sciences (KNAW) and the Dutch Research Council (NWO). ASML indicated that they were also open to other private partners, which was an important point for NWO: 'A future-proof institute must also have a healthy life cycle: parties can gradually join or leave, and the composition follows the development of the research themes'.⁶⁰ However, a condition for joining was that it could not result in either 'academic competition' or 'direct competition' with ASML.⁶¹ Newcomers were expected to adapt to the 'working style' and 'collaborative nature' of ARCNL.

Frenken's vision was quite similar to that expressed in the evaluation organised by KNAW and NWO in 2019.⁶² Although ARCNL was initiated by ASML, it primarily became an NWO institute. As such, the exclusive relationship with ASML was somewhat uncomfortable. Both in terms of research programme and partners, the critical evaluation committee believed that

ARCNL should broaden its scope to justify its position as an NWO institute.[63] This strengthened Frenken's resolve to follow the path he wanted to take. Meanwhile, ASML had been sending out mixed signals about expanding the consortium, and ASML board members were regularly stepping on the brakes. Frenken saw the evaluation as 'an extra incentive' to continue working on the expansion: '[The committee] has doubts about whether ARCNL has a right to exist if it does not more clearly take on a national role!'[64] The exclusive relationships with ASML and the universities in Amsterdam were seen as a weak point for the future of ARCNL.

Among the institute's direct partners, the initial scepticism had given way to enthusiasm. From 2019 onwards, both the scientific community and ASML began to show a much greater appreciation for ARCNL. In the eyes of the annual scientific evaluation committee, the institute had succeeded in combining blue-sky research with the industrial interests of ASML in a fruitful and harmonious way that was 'reminiscent of the model formerly underlying research centers such as the Bell Labs and the Philips Research Lab'.[65] This meant that ARCNL was even starting to become an model institute for them that could dispel existing cynicism about public-private partnerships. The interaction with ASML was felt to be increasingly 'natural', according to ARCNL group leader Paul Planken. Bart Noordam likewise confirmed that ARCNL was 'recognised more and more at ASML'.[66]

Nevertheless, the question of the type of science practised at ARCNL remained a topic open to different interpretations. In the new vision for ARCNL, roughly one-third of the research would be considered free space. Frenken interpreted this to mean that it could be used for research that was not yet on ASML's radar but had the potential to be in a few years. However, the scientific advisers interpreted this space as an opportunity to conduct research that could be completely independent of ASML or possibly linked to other external partners.[67] The committee also placed greater emphasis on high-impact publications, but Frenken felt this was incompatible with ARCNL's set-up. '[T]he application-driven (future) interest of ASML [...] cannot simultaneously be connected primarily with the orientation towards high-impact publications'. Although not 'mutually exclusive', this was only possible if part of the budget was set aside for 'unrestricted topics' about which 'high-risk, high-gain' research could be conducted 'without the simultaneous condition of alignment with ASML's needs and interests'. The scientific advisers' stance could only be maintained if they interpreted the use of the free space very liberally.[68]

Some saw the 'free space' as a means of steering ARCNL back towards academic research. The SAC saw that space as a 'nursery' for research that 'may or may not become of interest to ASML'. They would have preferred to see this leeway used explicitly as 'free space', particularly to give ARCNL a more prominent position in the academic world. Doing so could positively contribute to the institute's revenue-generating capacity, recruitment and potential partnerships. In their view, some group leaders did not feel comfortable enough to use the free space for free research. They told Frenken that he should be 'letting them roam a little bit more freely than has been the case up to now', which could also lead to 'significant benefits, not only for ARCNL, but likely for ASML as well'.[69] However, it was clear to Frenken that the free space also had to align with ARCNL's mission: it still had to have the potential to be relevant to ASML. It could not be a completely no-strings-attached domain.[70]

The ideal ARCNL researcher

The appreciation for ARCNL's research projects was closely related to the appreciation for the individual scientists and their characteristics. This was partly a practical issue: in 2019, ASML was very pleased with the 'continued inflow of highly qualified ARCNL employees'.[71] And in 2022, they developed a vision on the scientists at ARCNL. ASML greatly valued the groups working on metrology and the light source. They respected the 'independent manner' in which these researchers did their work, and especially respected that they functioned as 'sparring partners that dare to take [a] position in discussions'.[72] The culture at ASML fostered a respect for researchers who actively engaged with criticism rather than retreating from it. This was apparently less developed in the younger materials group, and ASML stated that they 'would like to encourage them to step up more and take independent positions, where they can really act as sparring partners and thought leaders'.[73] It was clear that being an ARCNL researcher required scientists to demonstrate certain other qualities.

Throughout the entire process of reflection surrounding the reform of ARCNL, much more explicit thought was given to the question of what type of scientist would thrive in the environment that had been created. What exactly was expected of an ARCNL researcher? This is what historians of science describe as scientific personae: our definition of 'a good scientist' is not determined purely by the quality of their experiments and theoretical insights, but also by the values and virtues they embody, such as objectivity and trustworthiness.[74]

What were the requirements for a researcher at ARCNL? What was specifically different about research there? Joost Frenken noted that research at ARCNL was a different kind of academic research: 'You don't give up your identity, but your identity becomes somewhat different'.[75] This applied not only to management and group leaders, but also to PhD candidates and postdocs. It was not only about 'skills', but also about 'character traits': the researchers needed a certain type of personality 'to get on well with the people at ASML'.[76] Above all, people had to be 'ASML-minded': collaboration with the company had to be seen as a challenge, not a burden.[77] The new ARCNL 2.0 set-up aimed for what Frenken later described as 'real interaction' in which the line of research was developed through very intensive consultation: 'You tell ASML what you think should happen and they respond to it [...]. It's a joint effort in which everyone is respected for who they are.'[78] Of particular interest to ASML was how the scientists took part 'in the game'.[79] They especially appreciated 'how someone could challenge someone else without irritating them'.[80] Starting in 2018, ARCNL group leaders had more intensive interaction with ASML: the collaboration became a fundamental part of the scientific work.

Prospective researchers were evaluated based on specific qualities. The recruitment of potential group leaders offers a glimpse into those desired qualities. First and foremost, they had to demonstrate flexibility around their research topic. It had to be clearly geared towards ARCNL: 'We have to face the fact that in this respect, ARCNL is more restrictive for group leaders in the freedom they have with their research program than a university department would be [...]'.[81] Personal qualities were also important for working with ASML. Potential group leaders were specifically assessed on their ability 'to run and manage the contacts with the application-driven researchers at ASML', especially when it came to experiment-focused positions.[82]

The new policy offered more space and flexibility to set up or phase out research, which in a sense could also be a way of selecting the desired type of researchers. Chair of the board Van Vuren agreed with Frenken's suggestion that one-third of ARCNL's budget be set aside for free research, but he stressed that this should not lead to 'endless [...] freewheeling'. This so-called breeding ground also had to be regularly 'purged' to reserve it mainly for 'scientific crown jewels with fresh ideas for excellent science'. He believed it could be a mechanism to send people 'who do not fit in with ARCNL in terms of drive or mentality back to a university or to a job elsewhere'.[83]

Scientific quality was not the only criterion in this selection procedure. Research groups that had obtained prestigious grants and received positive evaluations from both the university and industry could still be phased out. The entire research theme of processes, which included the groups for nanophotochemistry (Brouwer) and EUV photoresist (Castellanos), began to be phased out in 2018. This decision was not based on the quality of the research; instead, it was driven primarily by the theme's 'connection to ASML's research interests'.[84] So a good ARCNL scientist not only conducted quality research but also ensured it could be connected to ASML.

At the same time, a compromise was worked out to respect the academic time scale of the research. Research topics would be given five years to get up to speed, a period that would cover one generation of PhD candidates. If after five years the 'ASML resonance' was too weak, the research would be phased out.[85] Phasing out meant that there would be no follow-up funding from ARCNL's basic budget. In such a case, senior group members could count on 'stimulated emission': support for a transfer to another academic institute or industry.

While the above requirements mainly applied to group leaders, the nature of ARCNL was also reflected in the way PhD candidates progressed through their programmes. It was important that they engage in 'expectation management'. The SAC observed that it sometimes took PhD candidates several years to realise that ARCNL did not fully meet their expectations. In particular, they noted that 'academic freedom at ARCNL is more limited than they anticipated ahead of time, while the amount of time that they are asked to spend on (to some extent, classified) work that they cannot use for their thesis is more than expected or that they would have liked'.[86] That is why it was essential to create a clear set of expectations for PhD candidates.

At the same time, the position of PhD candidates and postdocs also raised questions about the course of study that best suited them. Instead of teaching, perhaps their time could be better spent on assignments, projects or consultancy for ASML: 'They can play an important role in knowledge transfer', said Stan Gielen of NWO.[87] In contrast, the academic partners thought that PhD candidates should become more 'immersed in the university community', because they 'seem to be confined in the ARCNL bubble'.[88]

Everything that applied to group leaders, PhD candidates and postdocs applied equally to the position of director. The ARCNL director had to operate in a complex field of influence. There were various stakeholders with high expectations, yet the institute itself was still developing.[89] The board saw the director as 'the leader' and the 'personification' of ARCNL.[90] A good director communicated openly, was able to delegate and make choices, set the course and encouraged collaboration between ASML and the universities in Amsterdam. Diplomacy was the director's most important quality: 'Identify opposing interests, but don't amplify them'.[91] The role was described as that of a 'travelling salesman' who had to build support for the research programme, both internally and among all the partners.[92] In addition, the director was the figurehead for the public-private partnership: a good director 'continues to promote the idea that basic research aimed at solving social and industrial problems combines the best of both worlds'.[93] ARCNL's leader needed to be 'supportive', 'inspiring' and 'solution driven'. So the board was unanimously enthusiastic about the reappointment of the director in 2018: with Frenken it was possible 'to reap in the coming years what had been sown in the past'.[94]

Conclusion

Starting in 2017, ARCNL underwent an intensive review process in which the relationships between the universities, ASML and ARCNL were redefined. While the relationships were increasingly seen in terms of conflicting views until 2017, the differences were embraced thereafter. The criticism from ASML was not resolved by keeping them at a distance, but rather by involving them much more closely in defining the research. A similar solution was found for the universities: they were not ignored but became more involved.

The period between 2018 and 2024 thus reflects ARCNL's steady-state phase: the institute had found its own way of working. This not only involved coordination at the board level but also affected the working methods and visions of science and scientists at the institute itself.

Notes

1. ARCNLVan Kolfschoten, 'Bot gedrag van chipmachinemakers leidt tot spanningen' ['Blunt behaviour of chipmakers leads to tensions'].
2. Steven Shapin, *The Scientific Life*, 229.
3. Joost Frenken & Udo Kock, 'Publiek-privaat onderzoek kan prima autonoom zijn' ['Public-private research can be perfectly autonomous'], *Het Financieele Dagblad*, 27 October 2021.
4. Jorrit Smit, 'Kennisoverdracht op de campus' ['Knowledge transfer on campus'], 119–143.
5. Van Eijck et al., 'De Campus-Economie; de bouwwoede van de universiteiten' ['The Campus Economy: The universities' building frenzy'], *De Groene Amsterdammer* vol. 139, no. 50 (9 December 2015).
6. 'About us', Matrix website, accessed 21 May 2024. https://www.matrixic.nl/en/about-us/; also see: 'Matrix; een onderdeel van Amsterdam Science Park' ['Matrix: part of Amsterdam Science Park'], *De Telegraaf*, 13 March 1990.
7. 'Instituutsmanager Marjan Fretz over nieuw pand ARCNL' ['Institute manager Marjan Fretz speaks about new ARCNL premises'], *Nieuwbrief Inside NWO-I*, March 2019. https://www.nwo-i.nl/artikel/instituutsmanager-marjan-fretz-over-nieuw-pand-ARCNL/
8. Ibid.
9. Jaap Huisman, 'Bijenkorf vol talent in gesloten bolwerk' ['A beehive full of talent in a self-contained stronghold'], *Het Parool*, 19 August 2019.
10. 'Instituutsmanager Marjan Fretz over nieuw pand ARCNL' ['Institute manager Marjan Fretz speaks about new ARCNL premises'], *Nieuwbrief Inside NWO-I*, March 2019.
11. Huisman, 'Bijenkorf vol talent in gesloten bolwerk' ['A beehive full of talent in a self-contained stronghold'].
12. Marc Laan, 'Willy Wortels in de wei' ['Mad scientists in the meadow'], *Het Parool*, 10 July 1999.
13. Buck Consultants International, *Inventarisatie en analyse campussen 2014* [*2014 inventory and analysis of campuses*], 35; 45. https://www.bciglobal.nl/uploads/9/20141138_definitief-rapport-analyse-campussen-nederland(1).pdf

14. Van Eijck et al., 'De Campus-Economie' ['The Campus Economy'].
15. Ibid.
16. Van Kolfschoten, 'Bot gedrag van chipmachinemakers leidt tot spanningen' ['Blunt behaviour of chipmakers leads to tensions'].
17. Joost Frenken, 'Response to SEP-report 2017', 23 November 2017. ARCNL archive; *Evaluation 2014-2016 ARCNL – Advanced Research Center for Nanolithography*, 14 November 2017, 28.
18. *ARCNL Evaluation 2014-2016 – Advanced Research Center for Nanolithography*, 14 November 2017, 31.
19. Ibid. 28.
20. 'Interim evaluation, away day and the like for ARCNL, For discussion at the ARCNL GB meeting on 11 May 2017', 25 April 2017. ARCNL archive.
21. 'Goals for reinvention of ARCNL, appendix to ARCNL GB Meeting on 11 May 2017'. ARCNL archive.
22. Joost Frenken, 'Questions to Stakeholders and ARCNL Staff', 14 June 2017. ARCNL archive.
23. *Blueprint of ARCNL 2.0* (Draft: confidential – 4 October 2017), appendix to Governing Board no. 48, 2 November 2017. ARCNL archive.
24. Ibid.
25. Planken, Castellanos Ortega, Versolato, 'Reaction from ARCNL staff to blueprint ARCNL 2.0', 25 October 2017. ARCNL archive.
26. Frank Schuurmans, 'ASML's response', 1 November 2017. ARCNL archive.
27. Ibid.
28. Stan Gielen to Van Vuren, Van den Hout, Hooijer, Visser, and Schouten, 'ARCNL', 12 October 2017. ARCNL archive.
29. 'Blueprint, perspectives of UvA and VU', 2 November 2017. ARCNL archive.
30. Stan Gielen, 'NWO opinion of the ARCNL 2.0 blueprint', 31 October 2017. ARCNL archive.
31. Response from Stan Gielen, 'RE: 'Draft ARCNL 2.0 blueprint', 1 October 2017. ARCNL archive.
32. Response from Joost Frenken, 'RE: 'Draft ARCNL 2.0 blueprint', 1 October 2017. ARCNL archive.
33. Notes from Hendrik van Vuren, meeting with Joost Frenken, 20 November 2017. ARCNL archive.

34. Message from Frenken to Hendrik van Vuren, 'Discussing the discussion', 21 November 2017. ARCNL archive.
35. Ibid.
36. Frank Schuurmans, with input from Jos Benschop and Wim van der Zander, 'ASML's response', 1 November 2017. ARCNL archive.
37. Van Kolfschoten, 'Bot gedrag van chipmachinemakers leidt tot spanningen' ['Blunt behaviour of chipmakers leads to tensions']; 'Kritiek van ASML; eind onderzoek' ['Criticism of ASML; research ends'], *Eindhovens Dagblad*, 7 September 2018.
38. *ASML ARCNL Evaluation Report 2017*. ARCNL archive.
39. Ibid.
40. Notes from Hendrik van Vuren, conversation with Joost Frenken, 28 November 2017. ARCNL archive.
41. *ARCNL SAC Report*, 2018. ARCNL archive.
42. Interview with Joost Frenken, 12 July 2024, Leiden.
43. Van Kolfschoten, 'Bot gedrag van chipmachinemakers leidt tot spanningen' ['Blunt behaviour of chipmakers leads to tensions'].
44. Email from Frenken to Stan Gielen, 'Strategic outline in conversation with stakeholders', 9 January 2018. ARCNL archive.
45. Correspondence between Frenken, Van Vuren and Fretz, 12 February 2018. ARCNL archive.
46. Interview with Joost Frenken, 12 July 2024, Leiden.
47. Email from Frenken to Stan Gielen, 'Strategic outline in conversation with stakeholders', 9 January 2018. ARCNL archive.
48. Ibid.
49. Ibid.
50. Joost Frenken to Gielen, Benschop, Maex, and Subramaniam, 'Towards a New ARCNL', 12 March 2018, ARCNL archive.
51. Ibid.
52. Ibid.
53. *ARCNL SAC Report*, 23-24 April 2018. ARCNL archive.
54. Ibid.
55. Ibid.
56. Ibid.
57. Message from Joost Frenken to Hendrik van Vuren, 18 May 2018. ARCNL archive.
58. Email from Van Vuren to Gielen, 'ARCNL business: ARCNL 2.0, Strategy Day, SAC Report, ASML Appreciation Report and reappointment of director', 18 June 2018. ARCNL archive.
59. Email from Van Vuren to ARCNL GB, 8 July 2018. ARCNL archive.

60. ARCNL Governing Board, report of meeting no. 53, 1 November 2018. ARCNL archive.
61. Memo from Joost Frenken, 'Guiding principles for the addition of new stakeholders to the ARCNL consortium, for the GB meeting on 25 January 2019', 11 January 2019. ARCNL archive.
62. *Rapport Portfolio-evaluatie* [*Portfolio evaluation report*], 10 January 2019, see: https://www.nwo.nl/sites/nwo/files/documents/Rapport%20Portfolio-evaluatie%2010%20januari%202019.pdf
63. Ibid., p. 17.
64. Memo from Joost Frenken, 'Guiding principles for the addition of new stakeholders to the ARCNL consortium', 6 March 2019. ARCNL archive.
65. *ARCNL SAC Report*, 2019, 28-29 April. ARCNL archive.
66. ARCNL Governing Board, report of meeting no. 56, 12 July 2019. ARCNL archive.
67. 'Response to the Report 2021 of the ARCNL Scientific Advisory Committee', appendix to ARCNL Year Plan 2021-2022 (draft), 25 June 2021. ARCNL archive.
68. Ibid.
69. *ARCNL SAC Report*, 2021. ARCNL archive.
70. ARCNL Governing Board, report of meeting no. 62, 9 July 2021. ARCNL archive.
71. ARCNL Year Plan 2020-2021, Responses to SAC & ASML Reports (draft), 1 July 2020. ARCNL archive.
72. *ASML Appreciation Report 2022*, May 2023. ARCNL archive.
73. Ibid.
74. Daston & Sibum, 'Introduction: Scientific personae and their histories', 1–8; Jeroen van Dongen & Herman Paul (Ed.), *Epistemic virtues in the sciences and the humanities* (Springer, 2017), 1–10.
75. Interview with Joost Frenken, 12 July 2024, Leiden.
76. Response from Van Vuren, 2 October 2017. ARCNL archive.
77. ARCNL Governing Board, report of meeting no. 50, 12 March 2018. ARCNL archive.
78. Interview with Joost Frenken, 12 July 2024, Leiden.
79. Interview with Jos Benschop, 4 June 2024, Veldhoven.
80. Ibid.
81. 'Advice Tenure-track GL Contact Dynamis', 27 June 2019. ARCNL archive.
82. Ibid.
83. Response from Van Vuren, 2 October 2017. ARCNL archive.

84. Memo from Marjan Fretz, 'Phasing out the processes research theme', 18 July 2018. ARCNL archive.
85. 'Towards a new ARCNL', 7 March 2018. ARCNL archive.
86. *ARCNL SAC Report*, 2019, 28-29 April. ARCNL archive.
87. Response from Stan Gielen, 'RE: 'Draft ARCNL 2.0 blueprint', 1 October 2017. ARCNL archive.
88. 'ARCNL SAC Report', 2020. ARCNL archive.
89. ARCNL GB memo, 'Points of interest for Joost Frenken on his reappointment as director of ARCNL for the 2019–2023 term', 22 June 2018. ARCNL archive.
90. Ibid.
91. Ibid.
92. 'GB interviews related to the reappointment of ARCNL's director', 13 June 2018. ARCNL archive.
93. ARCNL GB memo, 'Points of interest for Joost Frenken on his reappointment as director of ARCNL for the 2019–2023 term', 22 June 2018. ARCNL archive.
94. Hendrik van Vuren to the Works Council, 'Request for advice pursuant to Article 30 of the Dutch Works Council Act (WOR)', 22 June 2018. ARCNL archive.

⑤ Conclusion

Today, the Advanced Research Center for Nanolithography, a public-private research institute located at Amsterdam Science Park, is a concrete example of the growing importance of scientific research with recognisable social and economic impact. Its explicit mission is to conduct basic research inspired by the challenges faced by the Dutch high-tech industry, especially ASML. Now, in the year 2025, both the institute's stakeholders and outside observers recognise that it has succeeded in its ambitious mission. ARCNL's academic quality is underscored by a large number of successful research proposals and publications in prestigious international journals, and the group leaders have successfully built bridges to ASML.[1]

But ARCNL's path has not always been an easy one. Even before ARCNL was founded, some people at FOM raised questions about the institute's scientific relevance. In the early years, these were compounded by the financial preconditions set by the universities, for whom good science was partly about securing project funding. At the end of 2017, ASML's dissatisfaction also culminated in a situation that could be called a tipping point in the still-young history of the Advanced Research Center for Nanolithography. To a certain extent, the conflicts could be seen as administrative struggles, but at the same time they reflected different visions of what constitutes good science.

ARCNL brought together the worlds of academia, industry and science funding, which had to be coordinated at multiple levels. This book has divided this development into three phases, each of which also reflects a specific level of alignment. Between 2012 and 2013, the idea for ARCNL was born and developed by ASML, FOM, AMOLF and the universities in Amsterdam. Driven in part by the influence of the top sectors policy, the idea was quickly embraced by all parties involved. Starting in 2014, abstract visions were transformed into a tangible institute. As this process unfolded, it gradually became clear that the intense public-private partnership exposed tensions between the partners' expectations. With the revision in 2017 and 2018, it was recognised that the new institute could not simply represent a compromise between existing visions of science. It had to develop its own identity and vision.

Ultimately, the history of ARCNL sheds light on the process of institutional innovation. ARCNL's path was not a problem-free implementation of a preconceived concept, but rather an idea that evolved through twists and turns, peaks and troughs. When viewed as an institutional form of innovation, ARCNL refutes the now infamous 'linear model of innovation': the idea that a basic understanding can lead in a straight line to an application or implementable concept. In a sense, the same applies to ARCNL's scientific programme: the EUV technology is not only the end goal of technological development, but also the starting point for scientific research. In other words, it is not easy to determine where innovation begins and ends. This applies to both ARCNL as an institute and EUV technology.

Both ARCNL and the scientific programme relating to EUV technology show that innovation can best be understood in what historian Cyrus Mody has called a 'zigzag model' of innovation: a constant back and forth between theory and practice in which both people and ideas are in constant interaction. In a zigzag model of innovation, researchers constantly move between scientific and industrial environments. That leads them to develop 'a nuanced understanding of how to bounce between fundamental, curiosity-driven research and commercial, technological development [...]'.[2]

The history of ARCNL as presented in this book is more than just an institutional story. The set-up of this still-young institute has shed light on the way in which industry and academic science sought to draw closer to each other and the challenges this presented in practice. The partnership between ASML and Dutch universities touches on broader issues regarding science, innovation and the demands placed on the modern scientist. This final reflection therefore considers a few important lessons that can be learnt.

Moore's law in Dutch science

The founding of ARCNL overlapped with the arrival of the top sectors policy in the Netherlands, but this was not the sole reason for its creation. An important conclusion is that research institutes that collaborate with industry are not only the result of political choices but are equally shaped by the demands of industry. By 2012, ASML had already established many partnerships with Dutch academics. ASML's collaboration with FOM mainly took place via the FOM institute in Rijnhuizen. This changed in 2013 when Fred Bijkerk's optics group moved to the University of Twente. There, the group was represented as an 'industrial focus group' in the faculty.[3] At the same time, a FOM research group led by Hanneke Gelderblom was also set up as part of ASML. So both at universities and at ASML, there were existing institutional collaborations in which scientists built bridges between different worlds with apparent ease.

The public-private nature of ARCNL was primarily a response to the problems ASML was facing rather than a desire of the universities. The growth of the company and the need for more fundamental scientific knowledge led ASML to become increasingly visible in the academic world. Up to 2009, ASML had funded seven PhD candidates.[4] By 2015, the figure was said to be around 50 PhD candidates, and in 2023 ASML was funding approximately 150 PhD candidates, mainly on 'ad hoc projects'.[5] The creation of ARCNL was based on ASML's desire to work more intensively and on a longer-term basis with the academic sector. The high standards for EUV technology were embedded in a long-term research programme organised around metrology, the light source and materials research – the three major research themes of ARCNL.

The role ASML played in the creation of ARCNL shows the importance of the industrial perspective when we analyse science and innovation policy. The growth of public-private partnerships is often explained by the emergence of the idea that universities could be an engine of economic growth, a vision that focuses mainly on government policy. US historian Elizabeth Berman, for example, explained the emergence of 'university–industry research centers' in the United States in the 1980s as a result of 'government action' that encouraged a 'market logic' at universities.[6] Although government policy obviously played a large role in facilitating public-private partnerships, this is an incomplete explanatory model for the founding of an institute like ARCNL. Partnerships could be formed through individual PhD projects, large-scale consortia or concrete research centres. Industrial parties also ultimately determine the institutional form of public-private partnerships based on their own needs and desires.[7]

ASML was explicitly interested in a physical institute, not a virtual centre or consortium. For ASML, ARCNL was a means to study their themes and challenges on a larger scale in close collaboration with academic researchers. This also translated into their vision for the institute. With ARCNL, they 'opted for bricks and mortar'. To initiator Bart Noordam, an institute was not a virtual centre or consortium, but a physical building with windows, a door and a director.[8] This was a conscious choice on the part of ASML, 'because we believe in the importance of physical proximity, the chance encounters between different disciplines'.[9]

Science and scientists at the crossroads of academia and industry

Partnerships between industry and academia are often viewed with scepticism: does the 'sacred obligation' to seek the truth really fit with the profit motive of industry?[10] According to historian of science Steven Shapin, scientists became increasingly open to the idea of combining the two worlds, starting in the 1970s. To him, the entrepreneurial scientist was both 'a qualified scientist' and a 'risk taker', with 'one foot in the making of knowledge and the other in the making of artifacts, services, and, ultimately, money'.[11] He saw interactions between science and industry as social experiments 'in what motivates people and in how a range of motives – some of which have traditionally stood in conflict – might be satisfied together'.[12] Although Shapin's analysis focused mainly on individual scientists who founded companies, new scientific institutes are also, by definition, social experiments in which scientists have to contend with sometimes seemingly contradictory expectations. The combination of industrial and academic relevance is less self-evident at institutes like ARCNL than Shapin's view suggests. ARCNL developed its own scientific identity, and with it its own vision of its scientists.

Collaborating with companies was nothing new for many Dutch physicists. They had worked with companies such as Philips and Shell during the 20th century, and FOM had formally set up the Industrial Partnership Programme in 2004. Yet the academic world and that of ASML appeared to be light years apart in ARCNL's early years. Looking back, Joost Frenken realised that ASML's contribution was not simply that of funding for basic research – a view that he believes several researchers at ARCNL initially held. This attitude stemmed from the way academics had previously collaborated with companies: 'For us, a company was part of a mechanism to bring in grants'.[13] At the same time, ASML did not want contract research: 'They didn't want to keep us on a leash in the form of an entity that simply carries out assignments'.[14] ASML expected basic research from ARCNL, but Frenken noted that they 'were more critical than I was used to from other companies in terms of the spectrum of topics we were working on and the way we interacted with people at the company'.[15] Whereas consultations in usual public-private partnerships took place no more than twice a year, ASML wanted to work very closely with ARCNL.

The interaction between the researchers exposed the different frames of reference used by the Amsterdam scientists and ASML. In his book *Focus: The ASML Way*, journalist Marc Hijink describes ASML as a tough 'techie culture'. Technical discussions there are 'chaotic' and 'boisterous', and it is not uncommon for people to use the confrontation method in which colleagues openly challenge each other. Their focus is mainly on finding mistakes and less on celebrating successes.[16] In a farewell interview with *Inside NWO-I*, Frenken also noted that people at ASML 'generally don't applaud'.[17] The collaboration was unfamiliar territory for both parties in this respect. Previously, ASML had mainly collaborated with technical universities, and ASML's Senior Vice President Research Jos Benschop remarked afterwards that they 'play the game differently, they accept more from us'.[18] ASML had also learnt lessons from the collaboration. ARCNL researchers were all linked with ASML researchers, and Benschop noted that the latter were 'managing too closely' and 'far too aggressively'.[19] ARCNL took some getting used to for both the academics and the industrial researchers.

The partnership at ARCNL also touches on what historians of science refer to as scientific personae: values, qualities and character traits that shape the identity of the scientist. Qualities that are often seen as theoretical – like reliability, critical thinking and transparency – also manifest themselves at the individual level. In other words, the search for a new form of science implicitly involves a search for certain individuals with specific character traits. From a historical perspective, these values offer insight into the development of science in two ways. First, since the values change over time, they show that our understanding of good science depends on its historical context. Second, because different values rarely form a coherent whole, they offer insight into the sometimes seemingly contradictory expectations and influences that mould scientists.[20]

To build bridges between basic research and the world of ASML, they formulated a new vision of the scientist: an outline of the character traits and qualities that characterise the scientist at ARCNL. In a farewell interview, departing director Joost Frenken spoke about the 'courageous' group leaders who had come on board at ARCNL. The crisis around 2017 required 'manoeuvrability' and 'empathy', and the scientists were expected to have a 'genuine willingness' to listen to the expectations from industry. In Frenken's view, this dynamic was best suited to young researchers because they did not experience it as a loss of autonomy.[21] Appreciated group leaders were 'energetic' and 'proactive', which benefitted their interaction with ASML.[22] The qualities of the ideal scientist were explicitly defined in terms of personal characteristics, many of which emphasised social qualities that were also necessary to be able to do scientific work.

The history of ARCNL is thus part of a development of the values, qualities and virtues that were associated with scientists in the 20th century. The definition of the scientist was never set in stone and was not entirely in the control of the scientists themselves. What it means to be a scientist is defined in interaction with expectations from those around them. The 20th century saw a number of crucial developments that contributed to visions of science. As science funding organisations emerged, scientists had to master the skill of obtaining external funding; thus 'fundability' became an essential character trait of the modern scientist.[23] The advent of Big Science after the Second World War, with growing research groups, instruments and government funding, also exposed a tension between the moral and epistemic convictions of some scientists. Should science not remain as independent as possible from government and industry, and ideally be carried out by individuals?[24]

The development of scientists at ARCNL was admittedly on a much smaller scale than the aforementioned turning points, but it provides a glimpse into the tensions present in contemporary science. In particular, there is a tension between the values of industrial relevance and academic freedom, and neither concept is unambiguous. The idea of free and autonomous science is strongly associated with universities, but those researchers are limited by the strong focus on obtaining grants. In contrast,

ARCNL had the resources it needed to fund research, thanks in part to the support of ASML. Some scientists were willing to give up some of their freedom of choice if it meant more time could be freed up for research.

At the same time, concepts like relevance or impact are ambiguous. For example, research topics could not be too relevant to ASML because as soon as an ASML team became involved, 'this automatically means ARCNL is outnumbered and that there is short-term focus'.[25] In that case, ARCNL could no longer play a significant role in the research. However, when dealing with a topic that was not yet on ASML's radar, one had to consider the low level of interest and appreciation from ASML. In that situation, 'pushback [is] to be expected'.[26] Especially in this context, the character traits of the ARCNL researchers were important. A long-term vision 'requires a strong story and convincing arguments to gain the respect and understanding from ASML counterparts'.[27] Defining the impact and relevance of research is therefore also a social process in which researchers not only must have scientific knowledge, but also the persuasive power to engage in debate with ASML staff.

More so than an individual scientist who launches a start-up company after an academic career, ARCNL had to invent a genuine synthesis that could accommodate both academic and industrial aspects of research.

Notes

1. See, for example, the conclusions in the most recent SEP review: *Evaluation 2017–2022:* NWO *Institutes – ARCNL – Advanced Center for Nanolithography* (Academion, 2024).
2. Mody, *The long arm of Moore's law: Microelectronics and American science*, 224.
3. Egbert van Hattem, 'Onderzoek – Fred Bijkerk' ['Research – Fred Bijkerk'], *UT Nieuws* vol. 2, no. 3 (April 2013), 16–17.
4. Jos Benschop, 'ASML letter of support', 28 April 2009. NHA, FOM Archives, door 449, inv. no. 1566.

5. ASML-FOM Steering Committee, 3 November 2015. NHA, FOM Archives, door 449, inv. no. 1566; ARCNL Governing Board, report of meeting no. 67, 31 March 2023; ARCNL Governing Board, report of meeting no. 69, 10 November 2023. ARCNL archive.
6. Berman, *Creating the Market University*, 121.
7. Cyrus Mody, 'Academic centers and/as industrial consortia in American microelectronics research', *Management & Organizational History* vol. 12, no. 3 (2017), 285–303.
8. Interview with Bart Noordam, 26 June 2024, online.
9. Interview with Jos Benschop, 4 June 2024, Veldhoven.
10. Daniel S. Greenberg, *Science for sale: The perils, rewards, and delusions of campus capitalism* (University of Chicago Press, 2007). Also see: Chris Engberts, *Scholarly virtues in nineteenth-century sciences and humanities* (Springer, 2021), 207–217.
11. Shapin, *The Scientific Life*, 210.
12. Ibid., 264.
13. Interview with Joost Frenken, 12 July 2024, Leiden.
14. Ibid.
15. Ibid.
16. Hijink, *Focus*, 266–271.
17. Anita van Stel, 'Joost Frenken gaat van ARCNL naar de Rijksuniversiteit Groningen' ['Joost Frenken moves from ARCNL to the University of Groningen'], *Inside NWO-I newsletter,* June 2022. See: https://www.nwo-i.nl/artikel/interview-joost-frenken/
18. Interview with Jos Benschop, 4 June 2024, Veldhoven.
19. Ibid.
20. Van Dongen & Paul (Eds.), *Epistemic virtues in the sciences and the humanities*, 1–10; Kim M. Hajek, Sjang ten Hagen & Herman Paul, 'Objectivity, honesty, and integrity: How American scientists talked about their virtues, 1945–2000', *History of Science* vol. 62, no. 3 (2024), 442–469.
21. Anita van Stel, 'Joost Frenken gaat van ARCNL naar de Rijksuniversiteit Groningen' ['Joost Frenken moves from ARCNL to the University of Groningen'].
22. ASML *Appreciation Report 2016*, March 2017. ARCNL archive.
23. Noortje Jacobs & Pieter Huistra, 'Funding bodies and late modern science', *International Journal for History, Culture and Modernity* vol. 7 (2019), 887–898; Pieter Huistra & Kaat Wils, 'The exchange programme of the Belgian American Educational Foundation: An institutional perspective on scientific persona

formation (1920–1940)', *BMGN – Low Countries Historical Review* vol. 131, no. 4 (2016), 112–134.
24. Jessica Wang, '"Broken symmetry": Physics, aesthetics, and moral virtue in nuclear age America', in: Jeroen van Dongen & Herman Paul (Eds.), *Epistemic virtues in the sciences and the humanities*, 27–47.
25. 'Material department working group – summary and recommendations', 21 February 2022. ARCNL archive.
26. Ibid.
27. Ibid.

ARCNL at a glance

The Advanced Research Center for Nanolithography (ARCNL) is a research institute in Amsterdam that was founded in 2014 at the initiative of ASML, a major player in the world of semiconductor equipment, particularly lithography machines. With around 100 employees, including 70 researchers, ARCNL is a relatively small institute where people know each other and seek each other out for discussion and collaboration. ARCNL's mission is to conduct basic research that focuses on physics and chemistry-related challenges in nanolithographic technology. Its research programme is linked to the interests and challenges of ASML. ARCNL is one of the national research institutes under the umbrella of the Dutch Research Council (NWO) and is located at Amsterdam Science Park. The institute is a partnership with ASML, as well as with the two universities in Amsterdam and the University of Groningen.

ARCNL consists of 11 individual research groups that work together and share a common mission. In addition to a group leader, a research group is made up of technical staff and a variable number of PhD candidates who conduct research and prepare for a PhD in the natural sciences. As well as about 40–45 PhD candidates, ARCNL has about 20 postdoctoral researchers who gain further research experience for several years after obtaining their doctorate. ARCNL uses computer support, mechanical design support and a mechanical workshop.

ASML's most advanced lithography machines use extreme ultraviolet (EUV) technology. This involves a source of (light) radiation, specialised mechanical solutions to move wafers between sensitive measuring stations, and exposure stations where many microchips are 'written' on a single wafer using EUV light. The mirrors used to create the image are extremely advanced. Because EUV light is absorbed by the air, an EUV lithography machine is a very large vacuum chamber.

In addition to producing EUV light and the exposure, a lithography machine contains an incredible number of sensors, some of which have a metrological function: what does a wafer look like on the precision scale of nanometres and where exactly is it located? Metrology is also needed just outside the lithography machine to verify that the different layers that are written in the lithography machine are indeed directly above one another within a few nanometres. (One nanometre is $1/10,000^{th}$ of the thickness of a hair.)

The challenges outlined above form the foundation of ARCNL's basic research programme, which is divided into three broad themes.

- **Source:** Several groups in the source department are examining the process of generating EUV light. They study the details of this process both experimentally and using a modelling approach. The source department is looking for ways to improve knowledge about these sources and is examining alternatives to the current technological options.
- **Materials:** The materials department is conducting research into the behaviour of materials in nanolithography equipment. Materials wear out or are affected by EUV light radiation, so materials that are completely resistant to wear under normal circumstances can undergo changes inside the lithography equipment. The materials deteriorate because wafers are constantly placed on wafer tables, then clamped and removed. The challenge here is to clamp the wafers tightly without causing deformation and deterioration.
- **Metrology:** The third research department focuses on metrology. ASML often tests the limits of fundamental laws of physics. It is impossible to use visible light to display details that are much smaller than the wavelength of the light. This department uses a combination of techniques to examine how these limits can be circumvented. This could involve developing sources with a smaller colour (wavelength) so that smaller details can be 'seen'. Alternatively, the researcher could use knowledge of the different structures to focus on deviations from expected images. This department also looks at damage that can result from light used for metrology.

Acknowledgements

The research for *ASML and Dutch Physics* was carried out in 2024 at the University of Amsterdam and funded by the Advanced Research Center for Nanolithography as part of its tenth anniversary celebrations. The book is based on a literature review and archival research. Thanks to the openness and assistance of ARCNL, AMOLF, FOM and NWO, this research was able to rely on very recent source material. I am very grateful to all those involved. Conversations with people involved in ARCNL were an important guiding light in structuring the research; for this I thank Bart Noordam, Jos Benschop, Albert Polman, Huib Bakker, Hendrik van Vuren and Joost Frenken. I also thank Jeroen van Dongen, Wim van der Zande, Mario Daniels and Marc Assink for their involvement as the reading committee, as well as my colleagues from the History & Philosophy of Physics research group at the University of Amsterdam.

Archives consulted

ARCNL archive, Amsterdam

AMOLF archive, Amsterdam

National Archives, The Hague
Archive of the Dutch Research Council (NWO), access 2.25.109, inventory numbers 1691, 1669, 2326, 2336.

Noord-Hollands Archief, Haarlem
Archive of the Foundation for Fundamental Research on Matter (FOM), access 449, inventory number 1566.

Bibliography

Baggen, Peter, Jasper Faber & Ernst Homburg, 'Opkomst van een kennismaatschappij' ['Rise of a knowledge society'], in: Johan Schot et al. (Eds.), *Techniek in Nederland in de twintigste eeuw. Deel 7 [Engineering in the Netherlands in the 20th century. Part 7]* (Zutphen: Walburg Pers, 2003), 140–173.

Baneke, David, 'De veranderende bestuurscultuur in wetenschap en universiteit in de jaren zeventig en tachtig' ['The changing governance culture in science and academia in the 1970s and 1980s'], *BMGN – Low Countries Historical Review* vol. 129, no. 1 (2014), 25–54.

Baneke, David, 'De vette jaren: de Commissie-Casimir en het Nederlandse wetenschapsbeleid 1957–1970' ['The good years: the Casimir Commission and Dutch science policy from 1957–1970'], *Studium* vol. 5, no.2 (2012), 110–127.

Baneke, David, 'Organizing space: Dutch space science between astronomy, industry, and the government', in: Thomas Heinze & Richard Munch (Eds.), *Innovation in science and organizational renewal: Historical and sociological perspectives* (New York: Palgrave Macmillan, 2016), 183–209.

Baneke, David, 'Toegepaste natuurkunde aan de universiteit – contradictie of noodzaak?' ['Applied physics at university – contradiction or necessity?'], in L.J. Dorsman & P.J. Knegtmans (Eds.), *Universitaire vormingsidealen – de Nederlandse universiteiten sedert 1876 [University educational ideals – Dutch universities since 1876]* (Hilversum: Verloren, 2006), 29–38.

Banine, Vadime Y., EUV *lithography: Historical perspective and road ahead* (Eindhoven University of Technology, inaugural lecture, 2014).

Berman, Elizabeth, *Creating the Market University* (Princeton University Press, 2011).

Besselaar, Peter van den & Edwin Horlings, *Focus en massa in het wetenschappelijk onderzoek: de Nederlandse onderzoeksportfolio in internationaal perspectief [Focus and mass in scientific research: The Dutch research portfolio from an international perspective]* (The Hague: Rathenau Institute, 2010).

Bourzac, Katherina, 'A giant bid to etch tiny circuits', *Nature* vol. 487, no. 7408 (July 2012).

Broek-Honingh, Nelleke G. van den, L. Koens & A. Vennekens, *Totale Investeringen in Wetenschap en Innovatie 2018–2024 [Total investments in science and innovation from 2018–2024]* (The Hague: Rathenau Institute, 2020).

Browning, Larry & Judy Shetler, SEMATECH: *Saving the U.S. semiconductor industry* (Texas A&M University Press, 2000).

Chang, Hans & Dennis Dieks, 'The Dutch output of publications in physics,' *Research Policy* vol. 5, no. 4 (1976), 380–396.

Daniels, Mario & John Krige, *Knowledge regulation and national security in postwar America* (University of Chicago Press, 2022).

Dankbaar, Ben, 'Omgaan met de innovatieparadox. Bestaat er een kloof tussen universiteiten en bedrijven?' [Dealing with the innovation paradox. Is there a gap between universities and businesses?], *M&O: Tijdschrift voor Management en Organisatie* vol. 59, no. 1 (2005), 64–80.

Dao et al., 'NGL process and the role of International SEMATECH', *Proceedings Volume 4688, Emerging Lithographic Technologies VI* (2002), 29–35.

Daston, Lorraine & Otto Sibum, 'Introduction: Scientific personae and their histories', *Science in Context* vol. 16, no. 1/2 (2003), 1–8.

Delft, Dirk van et al., *Snaren, spiegels en plakband – 70 jaar Nederlandse natuurkunde* [*Strings, mirrors and sticky tape – 70 years of Dutch physics*] (W-Books, 2017).

Deuten, Jasper, *R&D goes global: Policy implications for the Netherlands as a knowledge region in a global perspective* (The Hague: Rathenau Institute, 2015).

Dongen, Jeroen van & Herman Paul (Eds.), *Epistemic virtues in the sciences and the humanities* (Springer, 2017).

Duijn, Jorijn van. *Fortunes of high-tech: A history of innovation at ASM International, 1958–2008* (Techwatch Books, 2019).

Engberts, Chris, *Scholarly virtues in nineteenth-century sciences and humanities* (Palgrave Macmillan, 2021).

Faasse, Patricia, *Profiel van een faculteit: De Utrechtse bètawetenschappen 1815–2011* [*Profile of a faculty: Sciences at Utrecht University 1815–2011*] (Hilversum: Verloren, 2012).

Flipse, Ab, 'Geen weelde, maar een offer, Vrije Universiteit, achterban en de natuurwetenschappen' ['Not opulence, but a sacrifice: VU Amsterdam, constituency and the natural sciences'], in: L.J. Dorsman & P.J. Knegtmans (Eds.), *Universiteit, publiek en politiek. Het aanzien van de Nederlandse universiteiten* [*Universities, the public and politics. The reputation of Dutch universities*] (Hilversum: Verloren, 2012).

Franzoni, Chiara & Francesco Lissoni, 'Academic entrepreneurs: Critical issues and lessons for Europe', in: Varga Attila (Ed.), *Universities, knowledge transfer and regional development: Geography, entrepreneurship and policy* (Cheltenham: Edward Elgar, 2009).

Georgescu, Iulia, 'Bringing back the golden days of Bell Labs', *Nature Reviews Physics* vol. 4, no. 2 (2022), 76–78.

Gertner, Jon, *The idea factory: Bell Labs and the great age of American innovation* (Penguin, 2012).

Gerven, Paul van & René Raaijmakers, *NatLab – Kraamkamer van ASML, NXP en de cd* [*NatLab: Breeding ground for ASML, NXP and the CD*] (Techwatch Books, 2016).

Greenberg, Daniel S., *Science for sale: The perils, rewards, and delusions of campus capitalism* (University of Chicago Press, 2007).

Haas, Wim de, K. van Assche, M. Pleijte & T. Selnes, *Gouden Driehoek? Discoursanalyse van het topsectorenbeleid* [*Golden triangle? A discourse analysis of the top sectors policy*] (Wageningen: Alterra University & Research Center, 2014).

Hajek, Kim M., Sjang ten Hagen & Herman Paul, 'Objectivity, honesty, and integrity: How American scientists talked about their virtues, 1945–2000', *History of Science* vol. 62, no. 3 (2024), 442–469.

Hijink, Marc, *Focus: De wereld van ASML. Het machtsspel om de meest complexe machine op aarde* [*Focus: The ASML Way. Inside the power struggle over the most complex machine on earth*] (Amsterdam: Balans, 2023).

Hoeneveld, Friso & Jeroen van Dongen, 'Out of a clear blue sky? FOM, the bomb, and the boost in Dutch physics funding after World War II', *Centaurus* vol. 55, no. 3 (2013), 264–293.

Hoeneveld, Friso, *Een vinger in de Amerikaanse pap: Fundamenteel fysisch en defensieonderzoek in Nederland tijdens de vroege Koude Oorlog* [*A finger in the American pie: Basic physics and defence research in the Netherlands during the early Cold War*] (PhD dissertation, Utrecht University, 2018).

Homburg, Ernst, *Speuren op de tast: een historische kijk op industriële en universitaire research* [*Navigating by instinct: A historical perspective on industrial and university research*] (Maastricht, 2003).

Huijen, Pim, 'Universiteit, bedrijfsleven en de opkomst van de beroepsonderzoeker 1880–1940' ['Universities, the business world and the rise of the professional researcher 1880–1940'], in L.J. Dorsman & P.J. Knegtmans (Eds.), *Onderzoek in opdracht. De publieke functie van het universitaire onderzoek in Nederland sedert 1876* [*Commissioned research: The public function of university research in the Netherlands since 1876*] (Hilversum: Verloren, 2007).

Huistra, Pieter & Kaat Wils, 'The exchange programme of the Belgian American Educational Foundation: An institutional perspective on scientific persona formation (1920–1940)', *BMGN – Low Countries Historical Review* vol. 131, no. 4 (2016), 112–134.

Jacobs, Noortje & Pieter Huistra, 'Funding bodies and late modern science', *International Journal for History, Culture and Modernity* vol. 7 (2019), 887–898.

Johnson, Ann, 'What if we wrote the history of science from the perspective of applied science?' *Historical Studies in the Natural Sciences* vol. 38, no. 4 (2008), 610–620.

Kersten, Albert, *Een organisatie van en voor onderzoekers: de Nederlandse organisatie voor zuiver-wetenschappelijk onderzoek (Z.W.O.) 1947–1988* [*An organisation by and for researchers: the Netherlands Organisation for Pure Scientific Research (ZWO) 1947–1988*] (Assen: Van Gorcum, 1996).

Kohler, Robert, *Partners in science: Foundations and natural scientists, 1900–1945* (University of Chicago Press, 1991).

Linden, Mowery, Ziedonis, 'National technology policy in global markets – Developing next-generation lithography in the semiconductor industry', in: Maryann Feldman & Albert Link (Eds.), *Innovation policy in the knowledge-based economy* (Springer, 2001).

Lintsen, Harry & Evert-Jan Velzing, *Onderzoekscoördinatie in de gouden driehoek. Een geschiedenis* [*Research coordination in the golden triangle: A history*] (The Hague: Rathenau Institute, 2012).

Mercelis, Joris, Gabriel Galvez-Behar & Anna Guagnini, 'Commercializing science: Nineteenth- and twentieth-century academic scientists as consultants, patentees, and entrepreneurs', *History and Technology* vol. 33, no. 1 (2017).

Miller, Chris, *Chip war: The fight for the world's most critical technology* (Simon & Schuster, 2022).

Mody, Cyrus, 'Academic centers and/as industrial consortia in American microelectronics research', *Management & Organizational History* vol. 12, no. 3 (2017), 285–303.

Mody, Cyrus, *The long arm of Moore's law: Microelectronics and American science* (MIT Press, 2016).

Noordam, Bart & Patricia Gosling, *Mastering your PhD: Survival and success in the doctoral years and beyond* (Springer, 2022 [2006]).

Noordam, Bart, *Het is tijd om promotieonderwijs en promotieonderzoek te organiseren* [*It is time to organise PhD education and research*] (Amsterdam: Vossiuspers, 2006).

Papaioannou, Dimitrios, *Multiphoton ionization of barium and sodium by intense picosecond laser radiation* (PhD dissertation, University of Viriginia, 1992).

Parikka, Jussi et al., *The Lab Book: Situated practices in media studies* (University of Minnesota Press, 2022).

Raaijmakers, René, *ASML's architects: The story of the engineers who shaped the world's most powerful chip machines* (Techwatch Books, 2018).

Raaijmakers, René, *De geldmachine – De turbulente jeugd van ASML* [*The cash machine – The turbulent youth of ASML*] (Techwatch Books, 2017).

Raizen, Mark, 'Commentary: Let's re-create Bell Labs!', *Physics Today* vol. 71, no. 10 (2018), 10–11.

Shapin, Steven, 'The Ivory Tower: The history of a figure of speech and its cultural uses', *The British Journal for the History of Science* vol. 45, no. 1 (2012), 1–27.

Shapin, Steven, *The Scientific Life: A Moral History of a Late Modern Vocation* (Chicago University Press, 2008).

Smit, Jorrit, 'Kennisoverdracht op de campus. Transferpunten, bedrijfscentra en science parks in de jaren tachtig' ['Knowledge transfer on campus. Transfer points, business centres and science parks in the 1980s'], in: Ab Flipse & Abel Streefland (Eds.), *De universitaire campus: Ruimtelijke transformaties van de Nederlandse universiteiten sedert 1945–2020* [*The university campus: Spatial transformations of Dutch universities from 1945–2020*] (Hilversum: Verloren, 2020), 119–143.

Smit, Jorrit, *Utility spots: Science policy, knowledge transfer and the politics of proximity* (PhD dissertation, Leiden University, 2021).

Streefland, Abel, *Jaap Kistemaker en uraniumverrijking in Nederland 1945–1962* [*Jaap Kistemaker and uranium enrichment in the Netherlands 1945–1962*] (Amsterdam: Prometheus, 2017).

Theunissen, Bert, *'Nut en nog eens nut'. Wetenschapsbeelden van Nederlandse natuuronderzoekers 1800-1900* [*'Usefulness and more usefulness': Visions of science of Dutch natural scientists from 1800–1900*] (Hilversum: Verloren, 2000).

Tjong Tjin Tai, Sue-Yen, J. van den Broek, T. Maas, T. Rep & J. Deuten. *Bedrijf zoekt universiteit – De opkomst van strategische publiek-private partnerships in* onderzoek [*Industry seeking university: The emergence of strategic public-private partnerships in research*] (The Hague: Rathenau Institute, 2018).

Toren, Jan Peter van den, L.K. Hessels, C. Eveleens en B.J.R. van der Meulen. *Coördinatie in de topsectoren. De geplande TKI's en hun uitdagingen* [*Coordination in the top sectors: The planned TKIs and their challenges*] (The Hague: Rathenau Institute, 2012).

Velzing, Evert-Jan, *Innovatiepolitiek: een reconstructie van het innovatiebeleid van het ministerie van Economische Zaken van 1976 tot en met 2010* [*Innovation politics: A reconstruction of the innovation policy of the Ministry of Economic Affairs from 1976 to 2010*] (Delft: Eburon, 2013).

VSNU, *Quality assessment of research – An analysis of physics in the Dutch universities in the nineties* (Utrecht, 1996).

Wadhwani, R. Daniel, Gabriel Galvez-Behar, Joris Mercelis & Anna Guagnini, 'Academic entrepreneurship and institutional change in historical perspective', *Management & Organizational History* vol. 12, no. 3 (2017).

Wang, Jessica, '"Broken symmetry": Physics, aesthetics, and moral virtue in nuclear age America', in: Jeroen van Dongen & Herman Paul (Eds.), *Epistemic virtues in the sciences and the humanities* (Springer, 2017), 27–47.

Witte, Pieter de, 'Public-private partnerships – an example from the Netherlands: The Industrial Partnership Programme', in: Wim Helwegen & Luca Escoffier (Eds.), *Nanotechnology Commercialization for Managers and Scientists* (2012), 263–290.

Reports

AWT, *Scherp aan de wind! Handvat voor een Europese strategie voor Nederlandse (top)sectoren* [*Full steam ahead! A guide to a European strategy for Dutch (top) sectors*] (2011).

ARCNL Self-Evaluation 2014–2016 (June 2017).

ASML Annual Report 2023.

ASML Annual Report 2003.

Evaluation 2014–2016 ARCNL – Advanced Research Center for Nanolithography (November 2017).

Evaluation report Amsterdam University Physics (February 2018).

Evaluation 2011–2016 AMOLF – Physics of Functional Complex Matter (February 2018).

Buck Consultants International, *Inventarisatie en analyse campussen 2014* [*2014 inventory and analysis of campuses*].

NWO, *Rapport Portfolio-evaluatie* [*Portfolio evaluation report*], 10 January 2019.

Evaluation 2017–2022: NWO Institutes – ARCNL – Advanced Center for Nanolithography (Academion, 2024).

Newspapers/online news

'Amsterdam wint strijd om instituut ASML' ['Amsterdam wins battle for ASML institute'], NRC Handelsblad, 27 May 2013.
'ASM Lithography Holding NV to acquire Silicon Valley Group Inc. in an all stock transaction valued at EUR 1.8 billion (US$1.6 billion)', ASML press release, 2 October 2000.
'ASM Lithography N.V. completes acquisition of Silicon Valley Group Inc'., ASML press release, 22 May 2001.
'ASML issues shares to TSMC in connection with Customer Co-Investment Program', ASML press release, 31 October 2012.
'ASML mag van Amerikanen SVG onder voorwaarden overnemen' ['Americans allow ASML to take over SVG under certain conditions'], de Volkskrant, 4 May 2001.
'ASML surpasses Nikon in '02 litho market', EE Times, 4 October 2003.
'Matrix; een onderdeel van Amsterdam Science Park' ['Matrix: part of Amsterdam Science Park'], De Telegraaf, 13 March 1990.
'Personalia', Het Financieele Dagblad, 18 July 2012.
Bijman, Reinier, Marianne Lamers, Roos Menkhorst & Tamar de Waal, 'Handelaar in bedrijfsgevoelige informatie' ['Trader in commercially sensitive information'], De Groene Amsterdammer no. 47, 23 November 2011.
Bregman, Rutger & Jesse Frederik, 'Maak kennis met de grootste uitvinder aller tijden' ['Meet the greatest inventor of all time'], De Correspondent, 25 February 2015.
Brugh, Marcel aan de, 'Van lappendeken naar waterhoofd' ['From patchwork to top-heavy'], NRC Handelsblad, 27 November 2014.
Calmthout, Martijn van, 'Kabinet snijdt in hoge promotiebonussen op universiteiten' ['Cabinet cuts high PhD bonuses at universities'], de Volkskrant, 25 November 2014.
Calmthout, Martijn van, 'Wondermateriaal van vuilniszakkenkwaliteit' ['Wonder material of bin-bag quality'], de Volkskrant, 29 June 2013.
Eijck, Guido van, Bram Logger, Saskia Naafs & Parcival Weijnen, 'De Campus-Economie; de bouwwoede van de universiteiten' ['The Campus Economy: The universities' building frenzy'], De Groene Amsterdammer vol. 139, no. 50 (9 December 2015).
Frenken, Joost & Udo Kock, 'Publiek-privaat onderzoek kan prima autonoom zijn' ['Public-private research can be perfectly autonomous'], Het Financieele Dagblad, 27 October 2021.
Funnekotter, Bart, 'Nederland is geen klein land in wetenschap; Haags beleid om onderzoek van universiteiten te sturen faalt, zegt het Rathenau Instituut' ['The Netherlands is not a small country in terms of science; The Hague's policy of directing university research is failing, says the Rathenau Institute'], NRC Handelsblad, 18 March 2011.

Giebels, Robert, 'Industriebeleid is verslechtering' ['Industrial policy is deteriorating'], *de Volkskrant*, 8 February 2011.

Giebels, Robert, 'Nederland gaat fundamenteel anders van kennis naar kassa' ['The Netherlands has a fundamentally different approach to how knowledge translates into cash'], *de Volkskrant*, 14 September 2011.

Graaf, Beatrice de, Appy Sluis & Peter-Paul Verbeek, 'Weer moet er geld naar die topsectoren in de wetenschap' ['Yet again, money is going to those top sectors in science'], NRC *Handelsblad*, 14 June 2011.

Grebenchtchikova, Anna, 'Topsectorenbeleid slecht voor kenniseconomie' ['Top sectors policy is bad for the knowledge economy'], *de Volkskrant*, 17 September 2012.

Hattem, Egbert van, 'Onderzoek – Fred Bijkerk' ['Research – Fred Bijkerk'], *UT Nieuws* vol. 2, no. 3 (April 2013), 16–17.

Holstein, William, 'U.S.-funded technology stays here, for now', *U.S. News & World Report* vol. 124, no. 19 (18 May 1998).

Huisman, Jaap, 'Bijenkorf vol talent in gesloten bolwerk' ['A beehive full of talent in a self-contained stronghold'], *Het Parool*, 19 August 2019.

Jongh, Hans de & Rutger Betlem, 'Grafeen, een wonderbaarlijk materiaal, lost grote belofte nog niet in' ['Graphene, a miracle material, does not yet fulfil its great promise'], *Het Financieele Dagblad*, 14 June 2014.

King, Ian & Cornelius Rahn, 'Intel investing $4.1 billion in ASML to speed production', *Bloomberg*, 10 July 2012.

Kolfschoten, Frank van, 'Bot gedrag van chipmachinemakers leidt tot spanningen' ['Blunt behaviour of chipmakers leads to tensions'], NRC *Handelsblad*, 6 September 2018.

Laan, Marc, 'Willy Wortels in de wei' ['Mad scientists in the meadow'], *Het Parool*, 10 July 1999.

Lammers, David, 'EUV gains as venture ends e-beam litho work', *EE Times*, 5 January 2001.

Lammers, David, 'U.S. gives ok to ASML on EUV effort', *EE Times*, 24 February 1999.

LaPedus, Mark, 'ASML refutes claim that SVG purchase will threaten EUV technology', *EE Times*, 17 April 2001.

Severt, Stef, 'Bedrijfsleven, niet overheid, moet innovatie stimuleren' ['Business, not the government, needs to stimulate innovation'], *Het Financieele Dagblad*, 26 October 2012.

Stel, Anita van, 'Joost Frenken gaat van ARCNL naar de Rijksuniversiteit Groningen' ['Joost Frenken moves from ARCNL to the University of Groningen'], *Inside NWO-I newsletter*, June 2022.

Verrijt, Harrie, 'Acute vragen ASML ook bij instituut' ['Burning questions for ASML also at institute'], *Eindhovens Dagblad*, 8 November 2013.

Verrijt, Harrie, 'Regio zette laag in op ASML-plan' ['Region placed low bid on ASML plan'], *Eindhovens Dagblad*, 29 May 2013.

Wennink, Peter, 'Creating value for all stakeholders', presentation at ASML Investor Day, 23 November 2014.